The geostationary applications satellite

CAMBRIDGE AEROSPACE SERIES

The geostationary applications satellite

PETER BERLIN

CAMBRIDGE UNIVERSITY PRESS

Cambridge

New York New Rochelle

Melbourne Sidney

PUBLISHED BY THE PRESS SYNDICATE OF THE UNIVERSITY OF CAMBRIDGE
The Pitt Building, Trumpington Street, Cambridge, United Kingdom

CAMBRIDGE UNIVERSITY PRESS
The Edinburgh Building, Cambridge CB2 2RU, UK
40 West 20th Street, New York NY 10011–4211, USA
477 Williamstown Road, Port Melbourne, VIC 3207, Australia
Ruiz de Alarcón 13, 28014 Madrid, Spain
Dock House, The Waterfront, Cape Town 8001, South Africa

http://www.cambridge.org

First published 1988
First paperback edition 2004

A catalogue record for this book is available from the British Library

Library of Congress cataloguing in publication data

Berlin, Peter.
The geostationary applications satellite/by Peter Berlin.
p. cm. - (Cambridge aerospace series)
Includes index.
1. Geostationary satellites. I. Title. II. Series.
TL796.6.E2B47 1988
629.43′4 - dc19 88-9560
CIP

ISBN 0 521 33525 6 hardback
ISBN 0 521 61603 4 paperback

Contents

In memory of
Monteraldo Schiavon
and
Don Baird

Preface

Twenty-two thousand miles above the equator, a very special family of man-made satelllites circles the earth. Basking in the sunshine, their wings dark blue and their bodies golden, they look like parrots perched side by side on an endless telephone wire. Most strain their ears to pick up messages from one part of the world and relay them to another. Some spend all their time observing the evolution of weather patterns in the atmosphere below. A few size up the earth to the nearest inch, while others perform scientific experiments. All of the satellites are hypochondriacal chatterboxes who mix tales about what they have just seen, heard or felt with frequent reports about their precarious health.

This is the family of *geostationary* satellites, so named because to an observer on the earth they appear to be fixed at one point in the sky. In fact they are not fixed at all but travel around the earth at the same rate as the earth turns about its axis. Unlike spacecraft in any other orbit, a geostationary satellite remains constantly within view of almost half the earth at all times, which is why it is so eminently suited for telecommunications and earth observation.

The spacecraft literature abounds with titles on *payloads*, such as telecommunication transponders, radiometers and scientific instruments. The rest of the spacecraft, called the *platform*, is usually only presented in outline, and the presentation of launch vehicles, orbits and programmatic issues is often schematic. The role of a payload is perhaps more glamorous than that of a platform, but from an engineering viewpoint the payload is merely the *primus inter pares*.

The purpose of the present book is to describe geostationary applications spacecraft technology from A to Z, taking an even-handed approach to launch vehicles, orbits, platforms, payloads and programmatic issues. Although the book concentrates on geostationary satellites, much of the text is also relevant to low-orbiting unmanned spacecraft. This is a vast range of subjects to cover in 214 pages, and inevitably the narrative has had to be condensed. Important topics such as military and scientific missions, ground stations, data links, control centres, data processing and operational management could therefore not be accommodated.

I have opted to show a cross-section of geostationary spacecraft technology as seen through the eyes of a project management team. Such a team has to acquire a broad perspective of technical progress in the context of performance, quality, schedule and cost. This perspective is lacking in the existing space engineering literature which largely consists of books on specialized subjects, compiled essays by several authors, and papers submitted to symposia; hence the present book which attempts to tell a coherent story about geostationary satellites.

In order to allow the reader to explore analytical issues in greater depth on his own, the text has been supplemented with basic mathematical equations which may be readily programmed on a personal computer.

The book is intended for undergraduate university students and for engineers and technicians associated with the space business. Lecturers and journalists, as well as management staff in industry and in space organizations, may also find it helpful.

My special thanks to go David Birdsall, Roger Moses and David Leverington for their critique of the substance and form of the typescript. I wish to express my gratitude to former colleagues of the European Space Agency for their valuable advice on specialized topics, and for their assistance in my search for literature references and illustrations. I am indebted to INMARSAT for granting me permission to write the book, and to members of the INMARSAT-2 satellite project team who patiently answered my barrage of questions on their subjects of expertise. Last but not least, I owe thanks to my wife Shirley for her steadfast encouragement, and for ironing out logical and editorial wrinkles in my writing.

Peter Berlin
May 1988

List of acronyms

Some of the most common acronyms in the spacecraft trade are listed below. Whenever a particular acronym has been used in this book, reference is given to the page where the arconym first appears.

		Page
ABM	Apogee Boost Motor	
AC	Alternating Current	100
AEF	Apogee Engine Fire	
AIT	Assembly, Integration and Test	201
AKM	Apogee Kick Motor	2
AM	Amplitude Modulation	163
AOCS	Attitude and Orbit Control Subsystem	199
ASE	Airborne Support Equipment (Shuttle)	16
ASW	Address & Synchronization Word	151
BAPTA	Bearing and Power Transmission Assembly	79
BER	Bit Error Rate	161
CFRP	Carbon Fibre Reinforced Plastic	75
CDR	Critical Design Review	
C/N	Carrier-to-Noise Ratio	161
DC	Direct Current	100
DOD	Depth of Discharge	98
DPA	Destructive Physical Analysis	190
EGSE	Electrical Ground Support Equipment	205

EIRP	Equivalent Isotropically Radiated Power	174
ELDO	European Launcher Development Organisation	24
EMC	Electromagnetic Compatibility	
EMI	Electromagnetic Interference	72
EOL	End of Life	
EPC	Electronic Power Conditioner	169
ESA	European Space Agency	151
ESD	Electrostatic Discharge	71
ET	External Tank (Shuttle)	
FDM	Frequency Division Multiplex	
FDMA	Frequency Division Multiple Access	166
FDR	Final Design Review	
FET	Field Effect Transistor	168
FIT	Failure in Ten-to-the-nine hours	192
FM	Frequency Modulation	163
FOV	Field of View	
FSK	Frequency Shift Keying	
GEO	Geostationary Orbit	2
GFRP	Glass Fibre Reinforced Plastic	
GMT	Greenwich Mean Time	50
GSO	Geostationary Orbit	
G/T	Gain-to-Noise Ratio	176
GTO	Geosynchronous Transfer Orbit	2
HPA	High-Power Amplifier	168
ICBM	Intercontinental Ballistic Missile	20
IF	Intermediate Frequency	167
I/F	Interface	
IR	Infrared	84
I_{sp}	Specific Impulse	5
ITO	Indium-Tin Oxide	85
IUS	Inertial Upper Stage	18
LEO	Low Earth Orbit	
LH	Liquid Hydrogen	
LNA	Low Noise Amplifier	168
LOX	Liquid Oxygen	
L/V	Launch Vehicle	

MGSE	Mechanical Ground Support Equipment	205
MLI	Multi-Layer Insulation	
MMH	Monomethylhydrazine	5
MRB	Material Review Board	196
NASA	National Aeronautics & Space Administration	78
NTO	Nitrogen Tetroxide	
OBDH	Onboard Data Handling	
OCOE	Overall Checkout Equipment	
OMS	Orbital Manoeuvring System (Shuttle)	13
OSR	Optical Solar Reflector	87
PA	Product Assurance	
PAM-D	Payload Assist Module - Delta	18
PCM	Pulse Code Modulation	154
PDR	Preliminary Design Review	
PFD	Power Flux Density	173
Pixel	Picture Element	183
PKM	Perigee Kick Motor	2
P/L	Payload	
PM	Phase Modulation	
PPL	Preferred Parts List	190
PSK	Phase Shift Keying	
PSS	Power Supply Subsystem	
PWM	Pulse Width Modulation	100
QA	Quality Assurance	195
RCS	Reaction Control Subsystem	
RCT	Reaction Control Thruster	
RF	Radio Frequency	159
RG	Rate Gyro	142
RIG	Rate Integrating Gyro	143
RX	Receiver	
S/C	Spacecraft	
SCOE	Special Checkout Equipment	
SCPC	Single Channel per Carrier	
SHF	Super High Frequency	
SIW	Spacecraft Identification Word	155

S/N	Signal-to-Noise Ratio	161
SPELDA	Structure Porteuse Externe de Lancement Double Ariane	10
SPF	Single Point Failure	
SRM	Solid Rocket Motor	13
S/S	Subsystem	
SSM	Second Surface Mirror	87
SSME	Space Shuttle Main Engines	13
STS	Space Transportation System (Space Shuttle)	13
SYLDA	Système de Lancement Double Ariane	10
TC	Telecommand	
TCS	Thermal Control Subsystem	
TDM	Time Division Multiplex	
TDMA	Time Division Multiple Access	166
TM	Telemetry	
TPA	Transistorized Power Amplifier	168
TT&C	Telemetry, Tracking and Command	149
TTC&M	Telemetry, Tracking, Command and Monitoring	
TWT	Travelling Wave Tube	169
TWTA	Travelling Wave Tube Amplifier	168
TX	Transmitter	
UDMH	Unsymmetrical Dimethylhydrazine	5
UHF	Ultra High Frequency	
VHF	Very High Frequency	
VIS	Visible	181

1

Launch Vehicles

Introduction

In early 1986, a series of launch vehicle disasters temporarily crippled space activities in the West. The first disaster was also the greatest when in January the space shuttle *Challenger* carrying seven astronauts and a very costly TDRS data relay satellite exploded 72 s after lift-off. In April a Titan vehicle blew up within seconds after lift-off, destroying a sophisticated military surveillance satellite; it was, moreover, the second successive loss of a Titan after a long history of successful launches. Shortly afterwards a Delta launch was aborted in flight when the main engine closed down prematurely due to an electrical fault. The US Weather Service lost an urgently needed GOES satellite in the process. As a direct consequence of the Delta accident, impending Atlas/Centaur launches were postponed for many months while design similarities between the two vehicles were investigated. Finally, in May, the European Ariane rocket brought down an Intelsat-VA communications satellite when the launch vehicle failed for the fourth time in only 18 launches.

In these circumstances, Western space organizations with satellites awaiting launch began to look eastward for substitute launch vehicles. The purpose of this initiative was twofold: to find launch opportunities in the medium term, and to forestall any attempt by a recovering Western launch service agency to establish a commercial monopoly. The USSR and China responded favourably with offers for flights on Proton and Long March, respectively. Although India and Japan were not yet in possession of rockets capable of launching medium- and large-size satellites, they were clearly making progress in achieving such capability. Given the launcher hiatus in the West, space agencies and news media obtained much more detailed design information about rockets in the East than might otherwise have been available. Today we are therefore in a position to study and compare a wider range of commercial launch vehicles than ever before.

The present chapter introduces fundamental rocket theory along with basic concepts of launch vehicle construction. The launch vehicles which will be used for geostationary satellite launches in the coming years are presented. The chapter concludes with a discussion on launch vehicle reliability.

Definitions

Before proceeding any further, let us define some basic launch-related concepts:

1. The *perigee* of an elliptic orbit is the point nearest to the earth.
2. The *apogee* is the opposite of the perigee, i.e. it is the point furthest away from the earth.
3. The *inclination* is an angular measure of the slope of an orbit with respect to the earth's equatorial plane.
4. The *launch trajectory* is the flight path of a rocket from lift-off until satellite injection into orbit.
5. A *parking orbit* is a low-altitude (150–300 km) circular satellite orbit.
6. A *geosynchronous transfer orbit* (GTO) is an elliptic satellite orbit with a perigee at 150–300 km above the earth and an apogee at geostationary altitude ($\simeq$ 35 800 km).
7. A *geostationary orbit* (GEO) is a circular satellite orbit at approximately 35 800 km altitude and with a 24-hr period; its inclination is 0°. A *geosynchronous orbit* is a 24-h orbit with arbitrary inclination and ellipticity.
8. A *perigee kick motor* (PKM) is an auxiliary rocket motor carried by a satellite to boost itself from a parking orbit into a GTO.
9. An *apogee kick motor* (AKM) is another auxiliary rocket motor needed to boost a satellite from GTO to GEO.
10. A rocket or satellite *propellant* is composed of a *fuel* and an *oxidizer*. If the two ingredients are stored together, they constitute a *monopropellant*; if stored separately, they form a

bipropellant. A *cryogenic* propellant is a normally gaseous agent that is rendered liquid through refrigeration.

Rocket Engine Architecture

Modern launch vehicles are equipped with either bipropellant liquid engines or solid propellant motors, or a combination of both.

The design of a typical *liquid engine* is shown in Fig. 1.1. The primary elements are the two propellant tanks and the thruster. One tank holds the fuel, the other contains the oxidizer. Turbopumps withdraw propellant from the tanks and feed it under pressure to the thrust chamber. Before entering the chamber, the propellant passes through injectors which help diffuse the liquids to a fine mist. Ignition follows as soon as the diffused fuel and oxidizer mix, and the resulting combustion yields the desired rocket thrust. If combustion occurs spontaneously,

Fig. 1.1. Block diagram of an open-cycle liquid propellant rocket engine.

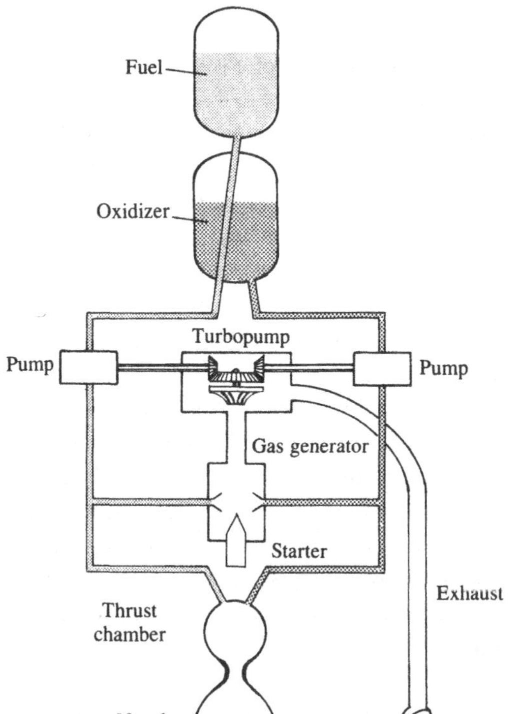

the propellants are said to be *hypergolic*. Non-hypergolic propellants have to be lit by an igniter.

The fuel and oxidizer pumps are driven by a turbine which runs on the same propellant as the rocket. Some engine designs have both pumps mounted on a common shaft which is either coupled directly to the turbine drive shaft or connected via a gearbox. Other engines employ independent turbopumps for the fuel and oxidizer. Combustion takes place in a separate gas generator which is a miniature version of the main thrust chamber. A charge of solid propellant inside the gas generator is ignited to start up the turbine.

Exhaust from the turbine is either released through a duct into the atmosphere (open cycle) or fed to the thrust chamber where it participates in the main combustion (closed cycle). Closed-cycle operation improves the efficiency of the engine but complicates the design.

Wheras liquid engines can be stopped and re-started in flight, *solid-propellant motors* burn continuously until all the propellant is consumed. They are used primarily where a high thrust level and simplicity are more important factors than operational flexibility, as in lift-off boosters and certain upper stages.

Chemical Composition of Propellants

A solid propellant is composed of a fuel, an oxidizer and a binding agent baked together (so-called *composite* propellant). Carboxy- or hydroxy-terminated polybutadiene with an aluminium powder additive is the most common fuel, while ammonium perchlorate (NH_4ClO_4) is the most frequently used oxidizer. The binder is usually a polymer which gives the propellant its characteristic rubbery consistency. The relative proportions of oxidizer, fuel and binder are approximately 70, 15 and 15% respectively.

Before a solid propellant motor is filled, the inside wall of the casing is coated with a rubber-like *liner* which improves the adhesion of propellant to the wall and serves as thermal insulation. The propellant is cast such that a contoured hollow core results. The core provides a larger burning surface than a solid casting, and a corresponding increase in thrust is obtained. By varying the dimensions of the hollow from one end to the other, the thrust profile may be tailored to the needs of a particular mission.

Most solid propellant motors use pyrotechnic igniters, though catalytic, hypergolic and electrical igniters are also available.

Liquid propellants may eventually phase out solids because of their superior-impulse and the possibility they offer to restart rockets during flight. *Monopropellants* burn by themselves after heating or catalytic decomposition, while combustion of *bipropellants* occurs when fuel is mixed with an oxidizer.

Some fuels used in launch vehicles and in satellite propulsion are as follows:

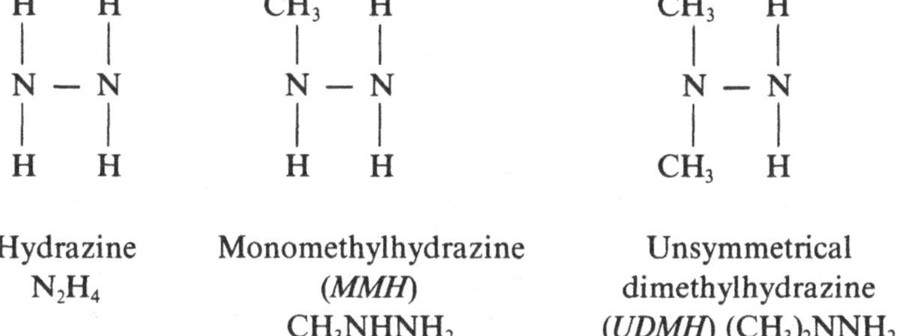

A form of kerosene known as RP-1 is used as a fuel in older launch vehicle designs. *Aerozine 50* is a 50/50 mixture of UDMH and hydrazine fuel. The latest vehicles employ cryogenic hydrogen (liquid H_2) for fuel together with cryogenic oxygen (liquid O_2) as oxidizer. "Cryogenic" means "very cold" and refers to the low temperatures required to maintain hydrogen and oxygen in liquid form during storage and filling (−253°C for H_2, −183°C for O_2). The most common non-cryogenic oxidizer is nitrogen tetroxide (N_2O_4).

Specific Impulse

A rocket engine imparts a net impulse on the launch vehicle. The *specific impulse* I_{sp} is defined as the exit velocity V_p of propellant exhaust through the engine nozzle divided by the earth's gravitational acceleration g. Alternatively, I_{sp} is the ratio between the thrust F/g and the rate dm/dt of expended propellant mass. Therefore, we have:

$$I_{sp} = \frac{V_p}{g} = \frac{F\,dt}{g\,dm}. \tag{1.1}$$

The specific impulse remains essentially constant during a burn and represents a convenient efficiency measure of individual rocket motor designs. The unit of specific impulse is *seconds*. Typical values of I_{sp} for the most commonly used propellants are:

	Typical I_{sp} (s)
Monopropellants	
Cold nitrogen gas	75
Solids	210–290
Hydrazine	220–300
Bipropellants	
Kerosene + O_2	250–290
$H_2 + O_2$	440–460
UDMH + N_2O_4	240–290
MMH + N_2O_4	300–315

The actual value within the I_{sp} range depends not only on the chosen propellant, but also on the rocket motor design and the combustion pressure. The higher the I_{sp}, the better the rocket performance in relation to mass. Because mass is nearly always a critical parameter in space flight, one might assume that the propellant with the highest specific impulse would invariably be adopted. There are other considerations, however, such as inherited (and therefore cheap and reliable) technology, manufacturing cost, operational convenience and safety.

The Rocket Formula

By examining Fig. 1.2, we may derive the all-important *rocket formula*. Take a rocket of mass m composed of propellant mass m_p and "dry" mass m_d (casing, nozzle, etc.). Assume that all the propellant mass m_p will be used up during a rocket burn, and that the mass is expelled with a velocity V_p relative to the rocket. Let the net rocket velocity gain be ΔV. Balancing the forces F gives us:

Fig. 1.2. Balance of momentum.

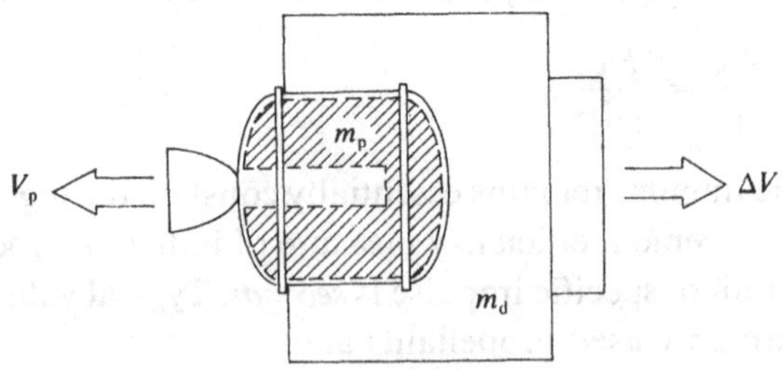

$$(F =)\, m\frac{\mathrm{d}V}{\mathrm{d}t} = -V_\mathrm{p}\frac{\mathrm{d}m}{\mathrm{d}t},$$

$$\mathrm{d}V = -V_\mathrm{p}\frac{\mathrm{d}m}{m},$$

$$\Delta V = \int_{d+p}^{d} \mathrm{d}V = -V_\mathrm{p}\int_{d+p}^{d} \frac{\mathrm{d}m}{m} = -V_\mathrm{p} \mathop{I}_{d+p}^{d} \ln m,$$

$$\Delta V = -V_\mathrm{p}\{\ln m_\mathrm{d} - \ln(m_\mathrm{d} + m_\mathrm{p})\},$$

$$\Delta V = I_\mathrm{sp} g \ln\left(1 + \frac{m_\mathrm{p}}{m_\mathrm{d}}\right). \quad (1.2)$$

Often some of the propellant remains onboard the rocket after a single burn, e.g. if there are several rocket stages, or if a restartable liquid engine is used. In this case the unburnt propellant should not be counted as a part of m_p but rather of m_d, because it is still a part of the ballast. To avoid confusion, it is safer mathematically to differentiate between the pre-burn mass m_1 ($= m_\mathrm{p} + m_\mathrm{d}$) and the post-burn mass m_2 ($= m_\mathrm{d}$ only if all the m_p has been spent), instead of using m_p and m_d. The difference between m_1 and m_2 is the mass of propellant actually expelled. Equation 1.2 then becomes:

$$\Delta V = I_\mathrm{sp} g \ln(m_1/m_2). \quad (1.3)$$

Up until the first satellite launch in 1957, most people imagined space rockets as single-stage devices which flew to the moon and back without shedding more than an occasional crew member en route to save mass. Nowadays, all launch vehicles are of a multi-stage design where burnt-out rocket stages are jettisoned before the next stage ignites. It seems obvious today that burnt-out stages constitute unnecessary ballast for the remaining rocket stages. This can be demonstrated mathematically as follows.

Let M and N be the stages of a two-stage rocket. Assume for simplicity that all the propellant is consumed during the burn of each stage, and that their I_sp is identical. $m_1 = m_\mathrm{p} + m_\mathrm{d}$ is the mass of the complete M-stage before burn, and $m_2 = m_\mathrm{d}$ is the mass after burn. Similarly, for the N-stage, $n_1 = n_\mathrm{p} + n_\mathrm{d}$, and $n_2 = n_\mathrm{d}$. From Eqn. 1.3:

$$\Delta V = I_\mathrm{sp} g \ln\frac{m_\mathrm{d} + m_\mathrm{p} + n_\mathrm{d} + n_\mathrm{p}}{m_\mathrm{d} + n_\mathrm{d} + n_\mathrm{p}} + I_\mathrm{sp} g \ln\frac{n_\mathrm{d} + n_\mathrm{p}}{n_\mathrm{d}}$$

or

$$\Delta V = I_\mathrm{sp} g \ln\frac{(m_\mathrm{d} + m_\mathrm{p} + n_\mathrm{d} + n_\mathrm{p})(n_\mathrm{d} + n_\mathrm{p})}{(m_\mathrm{d} + n_\mathrm{d} + n_\mathrm{p})\, n_\mathrm{d}}. \quad (1.4)$$

After some rearrangement of Eqn 1.4, we obtain:

$$\Delta V = I_{sp}g \ln \frac{m_1 + n_1}{m_2 + n_2} + I_{sp}g \ln \left(1 + \frac{m_d \cdot n_p}{n_d (m_d + n_d + n_p)}\right). \quad (1.5)$$

The first term to the right of the equal-sign in Eqn 1.5 represents the mass of the complete rocket if it had only comprised a single giant stage. The second term represents the additional ΔV made available by the two-stage configuration.

The Ascent Phase

Satellite designers refer to the powered flight phase from rocket ignition on the pad until final satellite separation as the *ascent phase.* Because satellites are exposed to violent accelerations, vibrations and shocks during this phase, their structural integrity must be assured through a judicious trade-off between stiffness and allowable mass. Other environmental factors, such as decompression and sun illumination, provide additional conditions for the spacecraft design.

During the early part of powered flight, satellites onboard an expendable launch vehicle are protected against aerodynamic friction by a *fairing*, also known as a *shroud.* The fairing is usually a straight or bullet-shaped nose cone encapsulating the satellite assembly and sometimes also the last rocket stage. Once the air density has dropped to a level where aerodynamic friction will no longer harm the satellites, the fairing opens up like the jaws of a crocodile, and the two halves are jettisoned to free the remaining rocket stages of the now unnecessary weight.

Fig. 1.3. Ariane launch trajectory.

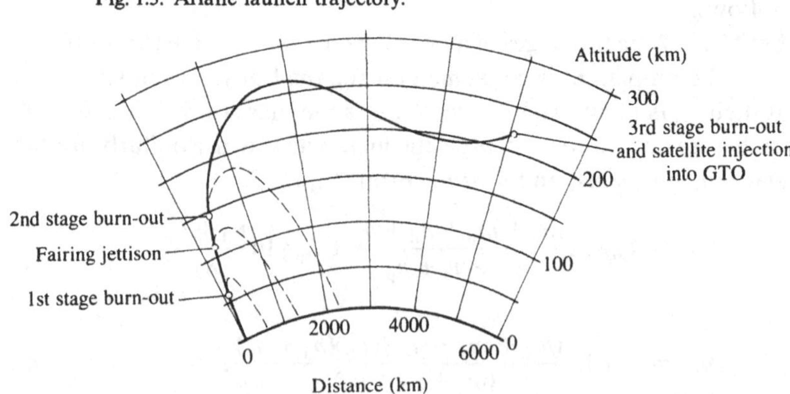

Early rocket designs such as the Scout brought satellites more or less directly into orbit, whereas Ariane and other contemporary launch vehicles take a more round-about path (Fig. 1.3). The reason for the latter is a need to optimize the performance of the rocket by minimizing the travel distance through the atmosphere and then gaining momentum through a nose-dive in relative vacuum.

Injection

When the last stage of the launch vehicle releases a satellite, we have *injection* of the satellite into orbit. In the case of geosynchronous satellites, the orbit following injection is either a parking orbit or a geosynchronous transfer orbit (Chapter 2). In order to reach geosynchronous orbit (Chapter 3), the spacecraft has to active its own propulsion subsystem (Chapter 9).

Expendable launch vehicles carrying more than one satellite into orbit stack them on top of each other with the use of a *tandem adapter*.

Fig. 1.4. Ariane SYLDA and SPELDA tandem adapters. Source: Arianespace.

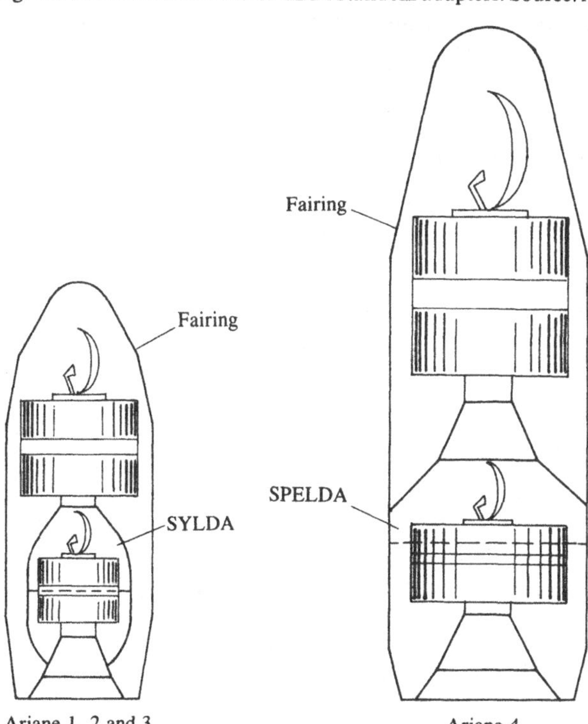

The purpose of the adapter is to allow the acceleration forces of each upper passenger to bypass rather than burden the passenger below. The SYLDA and SPELDA adapters available on Ariane are good examples (Fig. 1.4). The upper spacecraft is attached to the top of the adapter by a separation clampband. The lower passenger inside the adapter is also attached by a clampband at its base. When the last rocket stage has burnt out, the clampband holding the top-most passenger is released, and the spacecraft is ejected by compressed springs. The upper part of the adapter is then jettisoned, and the lower passenger follows suit shortly afterwards. The rocket stage is automatically re-oriented between separations to avoid collision between injected satellites and the adaptor cover. Manoeuvring also reduces the risk of the last rocket stage impacting on a released satellite. Strange as it may seem, rocket stages have been known to come alive again after nominal burn-out as residual fuel self-ignited (so-called *chugging*), and injected satellites have been severely damaged as a result (e.g. the radio amateurs' satellite OSCAR-10 following injection from Ariane in 1983).

Many launch vehicles spin up satellites before ejection either by employing a spin table equipped with roller bearings or by rotating the entire last stage. The spin motion provides gyroscopic stability (Chapter 10) while a satellite's attitude and orbit control subsystem is being switched on and checked out.

Launch Site Selection

The inclination i of an orbit is defined in Fig. 1.5. As long as the rocket travels in a planar trajectory without sideways deviations, i is a

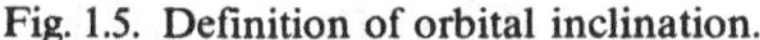
Fig. 1.5. Definition of orbital inclination.

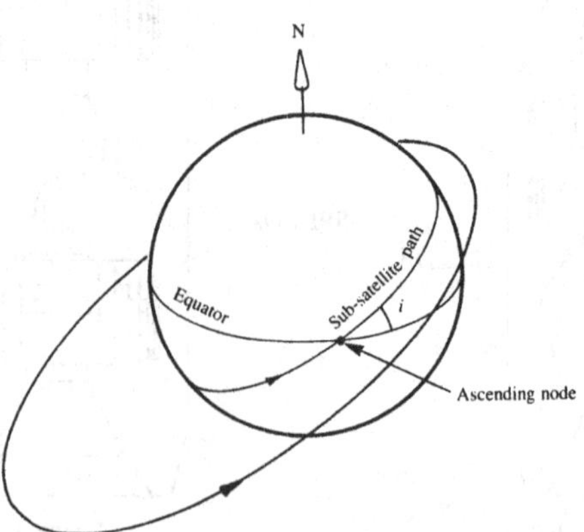

Fig. 1.6. Definition of launch azimuth *Az* and elevation *El*. The azimuth is measured clockwise from the North, while the elevation is measured upwards from the local horizon.

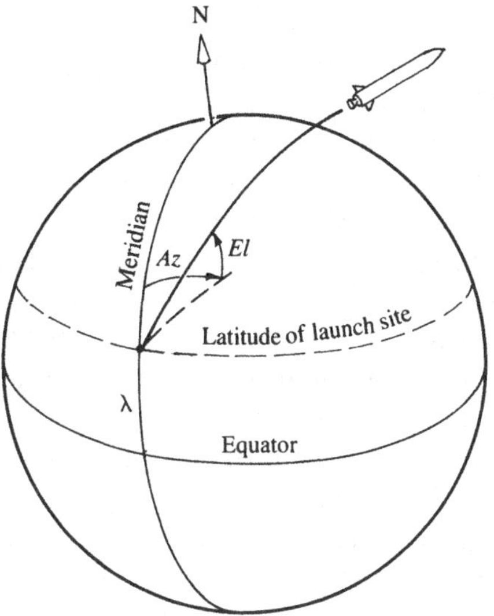

Fig. 1.7. A straight easterly (or westerly) launch gives minimum inclination i which is equal to the latitude λ of the launch site. Note that $i_6 = i_{\text{min}}$.

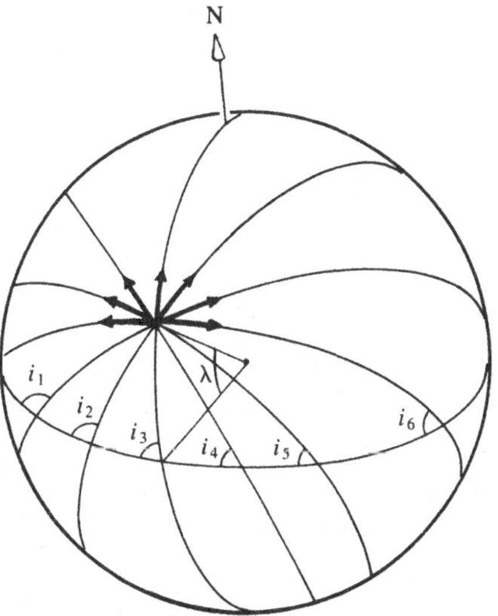

function of essentially two parameters, namely the latitude λ of the launch site and azimuth Az of launch (Fig. 1.6).

For $Az = 90°$ or $270°$, the inclination of a satellite transfer orbit immediately after injection will be equal to or greater than the latitude λ. Any other value of Az will result in $i > \lambda$ (Fig. 1.7). In order to get a free ride from the earth's rotation, satellites are nearly always launched in an easterly direction, i.e. with Az tending towards 90° rather than 270°.

The impossibility of achieving $i < \lambda$ for any Az poses problems for geosynchronous satellites which strive for inclinations near zero. To making things worse, it is usually not possible to launch in the ideal eastward direction because of safety problems associated with rockets overflying populated areas; hence $i > \lambda$ even at the best of times. The only solution is for either the launch vehicle or the satellite to undertake an out-of-plane *dog-leg manoeuvre* somewhere along the way to geosynchronous orbit. In some instances, the last stage of the launch vehicle performs a dog-leg manoeuvre during powered flight. In other cases satellites have to resort to their apogee motors to remove inclination in addition to circularizing the orbit.

The velocity change ΔV needed to obtain an inclination change Δi is obtained from Fig. 1.8 as:

$$\Delta V = (V_1^2 + V_2^2 - 2V_1V_2\cos\Delta i)^{1/2}. \quad (1.6)$$

From Eqn 1.3 we have:

$$\Delta m = m_1 - m_2 = m_1\left\{1 - \exp\left(-\frac{\Delta V}{g\,I_{sp}}\right)\right\}. \quad (1.7)$$

Fig. 1.8. Calculation of ΔV as a function of the plane change angle Δi.

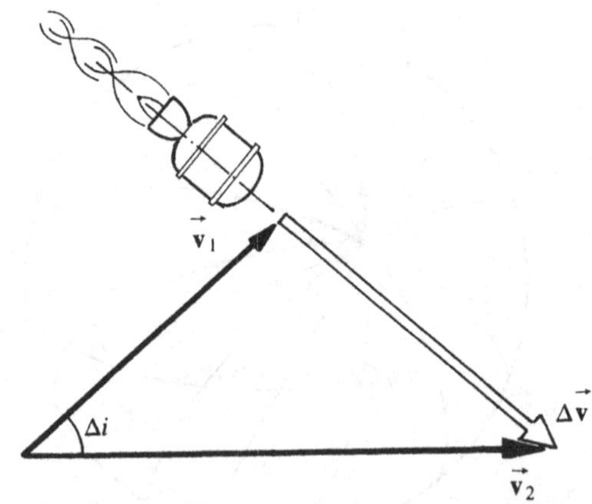

$$\Delta V = |\vec{v}_1 - \vec{v}_2| = \{(\vec{v}_1)^2 + (\vec{v}_2)^2 - 2\vec{v}_1 \circ \vec{v}_2\}^{1/2} = \{V_1^2 + V_2^2 - 2V_1V_2\cos\Delta i\}^{1/2}$$

By combining Eqns 1.6 and 1.7 we obtain a measure of propellant mass Δm needed to remove inclination. Dog-leg manoeuvres are wasteful in terms of propellant mass. This is why launch sites are built as close to the equator as possible within geographical, political and safety limits.

Note that there is no *upper* limit to the inclination achievable from a given launch site. A site at latitude λ can therefore launch satellites into any inclination from λ to $180 - \lambda$ simply by changing Az.

Description of Launch Vehicles

Seven vehicles are likely to be available for commercial launches of geosynchronous satellites in the foreseeable future. They are:

STS (Space Shuttle) (USA)
Titan III (USA)
Atlas G/Centaur 1-D (USA)
Delta II (USA)
Ariane-4 (Europe)
Long March 3 (China)
Proton SL-12 (USSR)

A condensed comparison of vehicle characteristics is given in Table 1.1.

Space Transportation System (STS)

Commonly known as the *Space shuttle*, the STS is composed of a reusable manned orbiter and two salvageable Solid Rocket Motors (SRM) attached to an external, expendable tank (Fig. 1.9). The tank provides cryogenic propellant (liquid H_2 and O_2) to the orbiter's three main rocket motors called the Space Shuttle Main Engines (SSME).

The orbiter carries the satellites and the crew. Its SSME and the two SRMs provide the necessary thrust during lift-off and early ascent. The SRMs are ejected 2 min into flight, leaving the orbiter and its external tank to continue the voyage into space. The tank is jettisoned 7 min later and is destroyed by aerodynamic friction during re-entry into the earth's atmosphere. The shuttle reaches its final 300-km circular parking orbit with the help of an Orbital Manoeuvring System (OMS) consisting of smaller liquid engines. The OMS is also used for de-orbiting before re-entry.

Table 1.1. *Characteristics of vehicles used to launch geostationary satellites*

	STS	Titan III	Atlas/ Centaur	Delta 3920/PAM	Ariane 4	Long March 3	Proton SL-12
Launch	NASA	Martin Marietta	General Dynamics	McDonnell Douglas	Arianespace	China, Great Wall Industrial Corporation	Glavcosmos
Launch site (for GTO)	Cape Canaveral	Cape Canaveral	Cape Canaveral	Cape Canaveral	Kourou, Fr. Guiana	Xichang	Baikonur
Overall height (m)	56	46	42	35	53–58	45	50
Lift-off weight (t)	1865	600	163	193	282 / 313	202	695
Booster propellant	Solid	Solid	Kerosene + O_2	Solid	Solid / UDMH + N_2O_4	—	—
Booster thrust (kN)	22 000	12 420	1679	3355	2700 / 3000	—	—
1st stage propel.	$H_2 + O_2$[a]	Aerozine + N_2O_4	Kerosene + O_2	Kerosene + O_2	UDMH + N_2O_4	UDMH + N_2O_4	UDMH + N_2O_4
1st stage thrust (kN)	6300	2429	269	921	3000	2750	9810
2nd stage propel.	[b]	Aerozine + N_2O_4	$H_2 + O_2$	UDMH + N_2O_4	UDMH + N_2O_4	UDMH + N_2O_4	UDMH + N_2O_4
2nd stage thrust (kN)	[b]	463	147	41	785	?	2400

3rd stage propel.	—	?[c]	—	Solid	$H_2 + O_2$	$H_2 + O_2$	UDMH + N_2O_4
3rd stage thrust (kN)	—	?[c]	—	?[d]	63	?	600
4th stage propel.	—	—	—	—	—	—	Kerosene + O_2
4th stage thrust (kN)	—	—	—	—	—	—	85
Standard GTO perigee height (km)	300	160	167	185	200	200	35 800[e]
Standard GTO inclination (deg)	26.5	28.5	28.5	28.5	7	31.1	0[e]
Payload capability to GTO (kg) at standard perigee and inclination	PAM-D: 1247 PAM-D2: 1842	Transtage 1977 PAM-D2: 1882	2200	1270 1430[g]	1900–4200[f]	1400	2000[e]

[a]Orbiter main engines.
[b]A variety of solid and liquid propellant perigee motors are available.
[c]Transtage, PAM-D or PAM-D2.
[d]PAM-D.
[e]Injection directly into GEO.
[f]Depending on number and type of boosters.
[g]Delta II (6925).

Up to four medium-sized geostationary satellites can be readily accomodated in the cargo bay of the orbiter. Most of those utilizing separate perigee motors (see below) are kept in cocoons called Airborne Support Equipment (ASE), which protect them from solar heating, allow last-moment battery charging and telemetry monitoring, and provide spin-up for stabilization purposes.

Upper stages

The low altitude and highly inclined shuttle parking orbit bears little resemblance to the 35 800-km equatorial geostationary orbit. All shuttle-launched geostationary satellites therefore need supplementary rocket motors, or *upper stages*, in the form of a *perigee kick motor* (PKM) and an *apogee kick motor* (AKM). The PKM is fired soon after the satellite has been released into space by the ASE. The impulse provided by the PKM sends the satellite into an elliptic geosynchronous transfer

Fig. 1.9. The Space Shuttle in orbital flight, configured for satellite release from the PAM-D2 cradle. The dash–dot contour shows the external tank and the solid rocket motors before jettison.

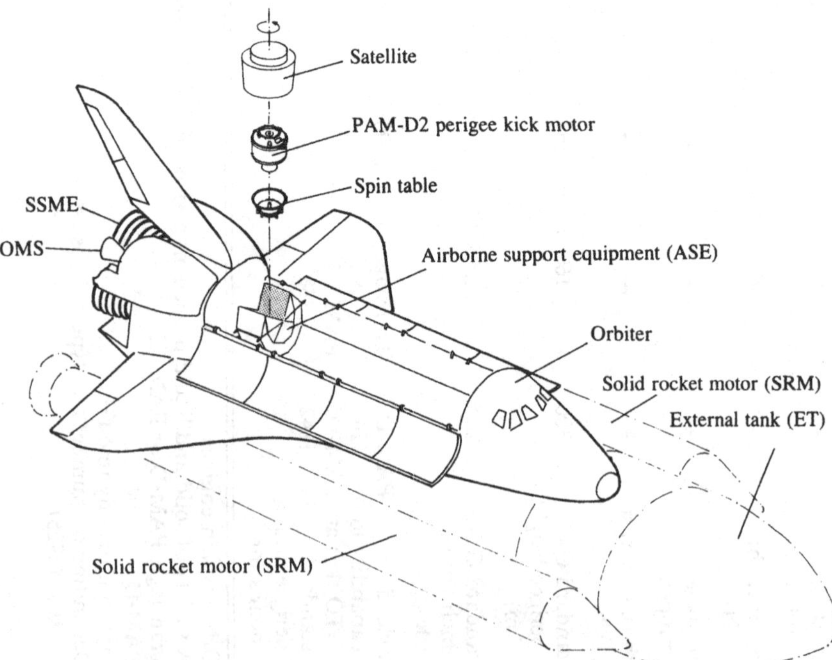

Fig. 1.10. Transition from parking orbit through GTO to GEO using a PKM and a *solid* propellant AKM.

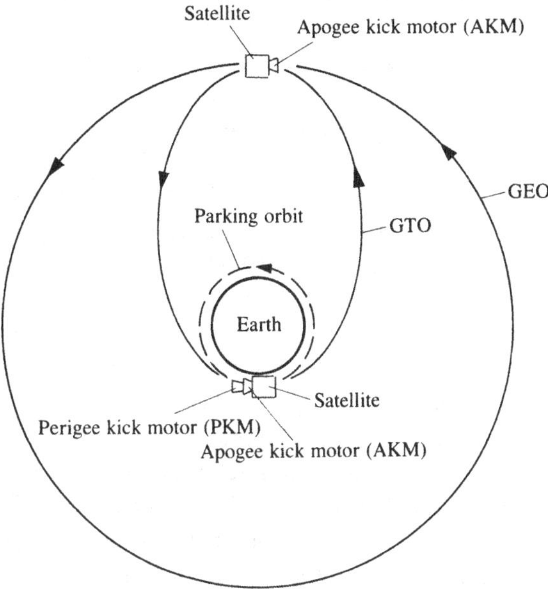

Fig. 1.11. Transition from parking orbit through multiple GTO to GEO using a PKM and a *liquid* propellant AKM.

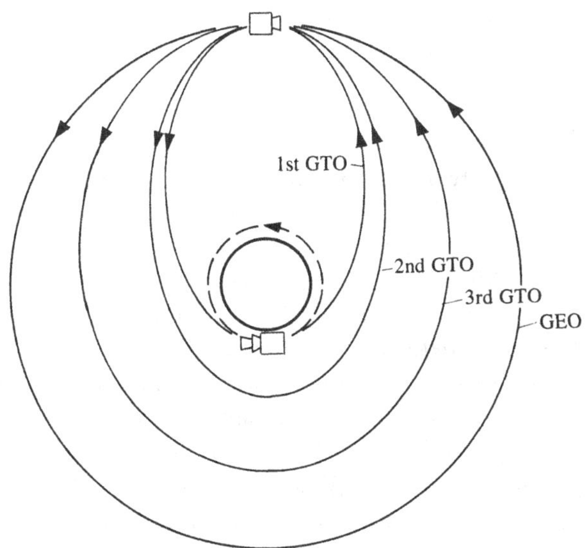

orbit (GTO), while the AKM circularizes the orbit at geostationary altitude (GEO) and removes most of the orbital inclination (Figs 1.10 and 1.11).

The best-known perigee motors are the solid-propellant Payload Assist Module (PAM-D) and its more powerful cousin PAM-D2. The motor is jettisoned after burn-out. The Inertial Upper Stage (IUS) is even more powerful and consists of two solid propellant rockets in tandem which combine the PKM and AKM functions in a single assembly. Other upper-stage designs are under development.

The term "upper stage" denotes a PKM, with or without an associated AKM, which is auxiliary to the satellite and which is jettisoned after burn-out. Most satellites actually include an AKM as an integral part of the design, and a few are equipped with an integral PKM as well. Such spacecraft do not require an upper stage for launch on the shuttle. AKMs are discussed further in Chapter 9.

Because satellites must carry so much rocketry of their own to reach geostationary orbit, the STS is not an ideal launch vehicle for such high-altitude missions. The main reasons for the shuttle's popularity have been the low launch price and the high reliability of the shuttle itself. The good reliability record has been marred, however, by PKM failures which have left satellites stranded in useless orbits.

STS launches for geostationary missions take place from Launch Complexes 39A and 39B at Cape Canaveral in Florida.

Titan III

Titan started out as an intercontinental ballistic missile (ICBM) in the 1950s. As the years went by, it flew interplanetary missions for NASA and a variety of military missions for the US Air Force. Titan III is the version currently proposed for commercial geosynchronous satellite launches. It is basically a two-stage vehicle augmented by solid boosters and with optional upper stages.

At first glance (Fig. 1.12) the vehicle resembles the shuttle without the orbiter. Two large SRMs are attached to the Aerozine/N_2O_4 first stage. The second stage uses the same propellant as the first. For the third stage, satellite customers have a choice of PKMs (e.g. the solid propellant PAM-D or PAM-D2) as well as the liquid propellant Transtage. Two satellites, each equipped with a PAM-D or PAM-D2 perigee motor, can be carried into a 200 km circular parking orbit. If they prefer to fly without their own perigee motors, then the Transtage will take them directly to GTO. Other upper-stage alternatives are under study.

A flight sequence begins with ignition of the two SRMs. These burn out 2 min after lift-off, at which time the first stage liquid engine is started. During the first stage burn the fairing is jettisoned. The second stage takes the satellites into a low, circular parking orbit. The upper stages (PKM and AKM) complete the launch sequence into GTO and GEO.

Titan III vehicles carrying geostationary satellites lift off from Launch Complex 40 and 41 at Cape Canaveral.

Fig. 1.12. The Titan-III launch vehicle with two satellites mounted in tandem.

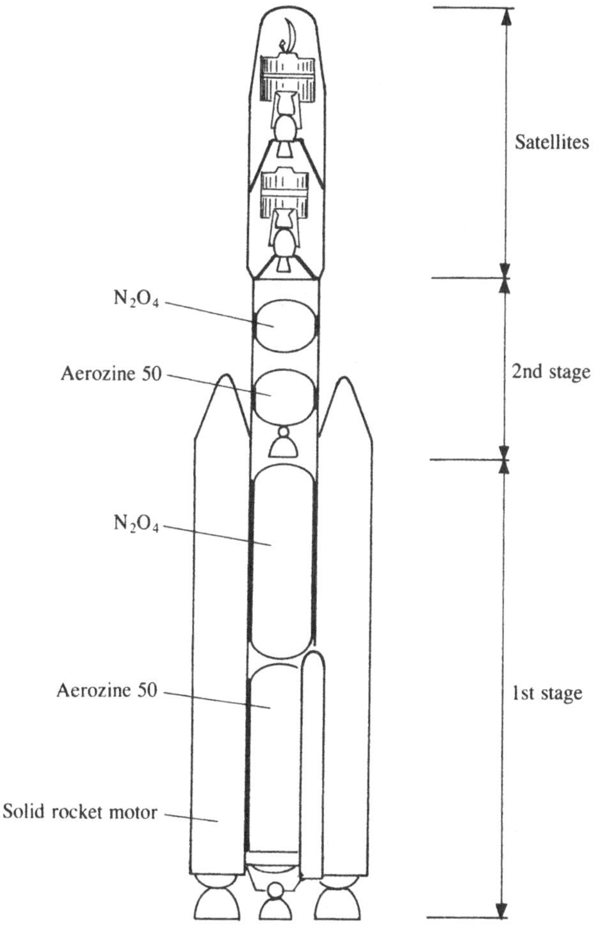

Atlas G/Centaur D-1

The Atlas/Centaur vehicle is the result of a marriage between two workhorses which at certain times have gone their separate ways. Like the Titan, the Atlas rocket started out as an ICBM in the late 1950s; it also launched astronauts into space in the Mercury–Atlas configuration of the early 1960s. The Centaur has taken time off from Atlas on some

Fig. 1.13. The Atlas/Centaur launch vehicle with a single satellite passenger.

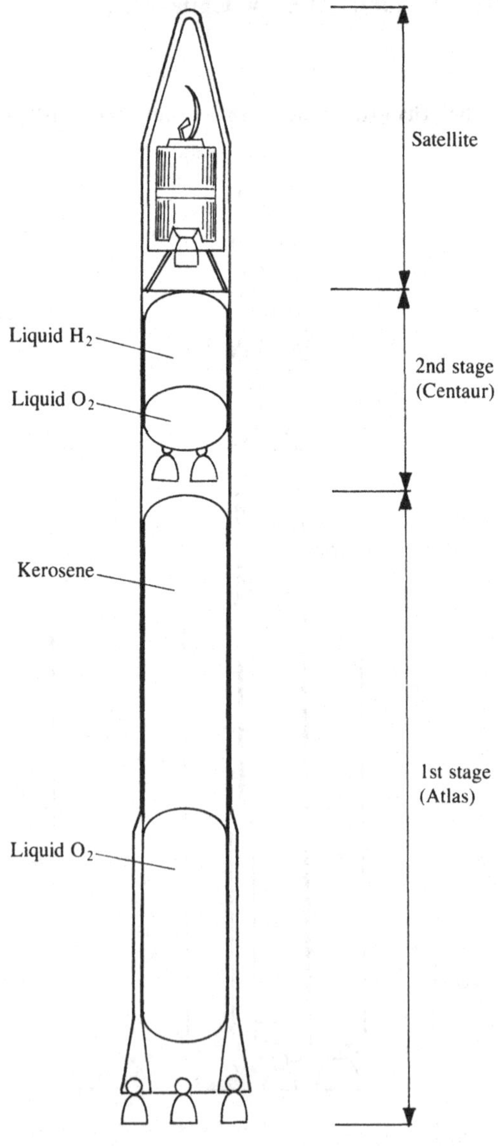

occasions to fly with Titan and was, until recently, proposed as an upper stage for shuttle-launched interplanetary payloads. The Atlas G/Centaur D-1 is the version which is currently being contemplated for commercial launches (Fig. 1.13).

The rocket is classified as a "two-and-a-half stage" vehicle because of its somewhat peculiar build. The Atlas first stage is equipped with a main *sustainer engine* and two auxiliary *booster engines*, all of which feed off a common kerosene/O_2 tank. The boosters are mounted on a skirt which is ejected 2.5 min into flight, leaving the sustainer engine to carry on alone. After separation from the expended Atlas stage, the cryogenic H_2/O_2 Centaur second stage takes over and executes two burns separated by a 15-min coasting phase. The last burn places the satellite in GTO.

Atlas G/Centaur D-1 vehicles for geostationary missions are launched from Launch Complex 36B at Cape Canaveral.

Delta II

The Delta is almost as old as its bigger cousins Titan and Atlas/Centaur. Nowadays it is a three-stage vehicle with up to nine solid-fuel strap-on boosters attached to the first stage. The Delta II rocket is an augmented version of the well-known Delta 3900 series (Fig. 1.14). The propellant used for the first stage is kerosene/O_2, whereas the second stage employs aerozine and N_2O_4. A slightly modified PAM-D is used for the third stage.

At lift-off, six of the nine boosters are ignited together with the first stage engine. Then, 1 min later the six are jettisoned and the remaining three take over for another 1 min. The main engine then completes the first part of the trajectory alone. During the second stage burn the fairing is released. After second stage burn-out, the satellite/PAM assembly is spun up to 30–120 r.p.m. on top of a spin table before separation. A brief coast phase follows before PAM ignition. At the end of the PAM burn, the satellite is injected into transfer orbit.

Each Delta is capable of lifting a single small- or medium-sized satellite into transfer orbit. Single-payload *dedicated* launches offer satellite customers considerable operational flexibility compared with multiple-payload *shared* launches but, not surprisingly, the launch cost is also somewhat higher. Yet more powerful versions of Delta II are under development.

Geostationary satellites are launched by Deltas from Complexes 17A and 17B at Cape Canaveral.

Ariane-4

Of all the expendable launch vehicles described in this chapter, Ariane is the only one designed exclusively for satellite launches from the outset. All the others have a heritage from ICBMs.

Fig. 1.14. The Delta 3920 launch vehicle with a PAM third stage and a single satellite passenger. Two of the nine strap-on boosters are shown.

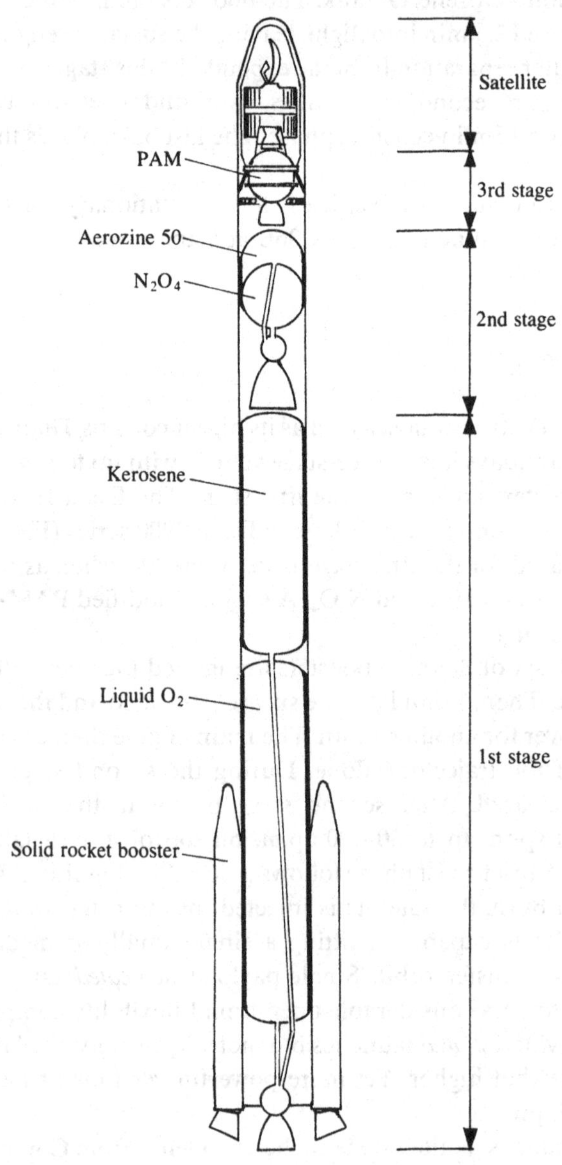

Fig. 1.15. The Ariane-4 launch vehicle with two satellite passengers using SPELDA. The version shown has two liquid and two solid propellant strap-on boosters.

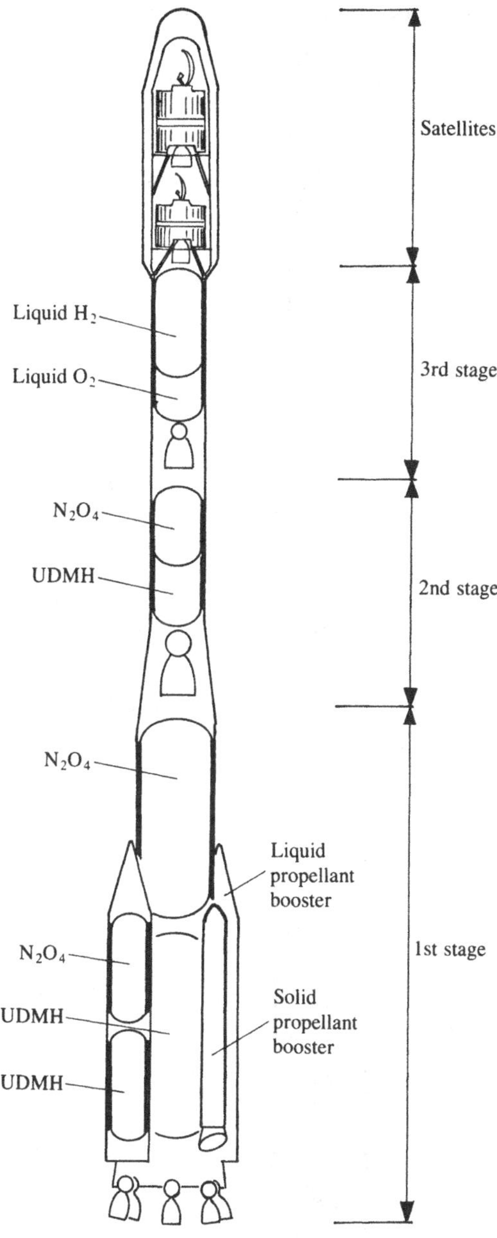

Ariane rose like a phoenix out of the ashes of the European ELDO rockets of the 1960s which never did manage to place a satellite in orbit. Like the shuttle in the US, Ariane has greatly contributed to the growth

Fig. 1.16. The Long March 3 launch vehicle with a single satellite passenger.

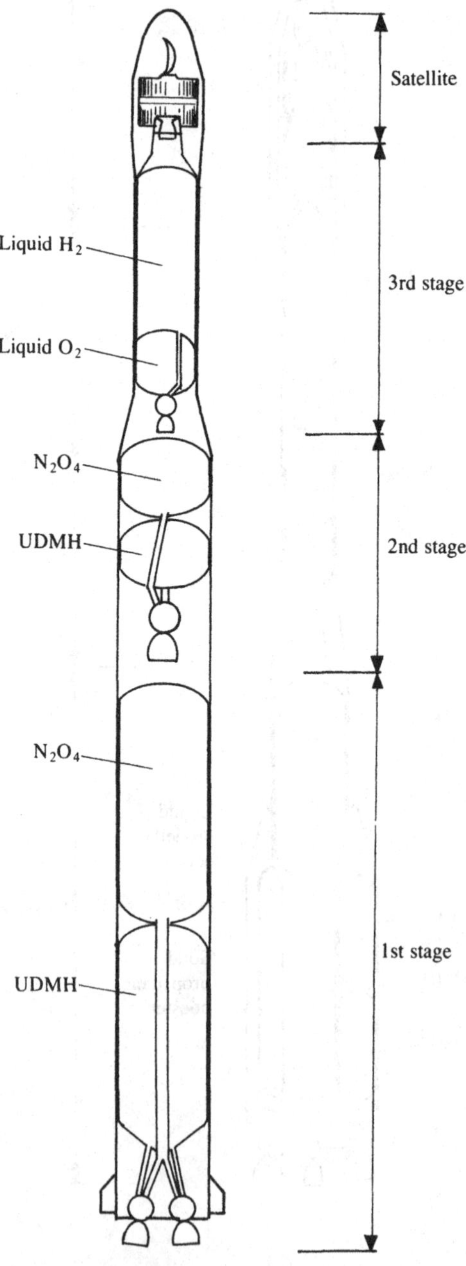

of Europe's technological self-confidence. Between 1981 and 1986 – the golden years for geostationary launches – the shuttle and Ariane dominated the launch service market in the West.

Ariane-1, the original version, made its maiden flight in 1979. It was a

Fig. 1.17. The Proton SL-12 launch vehicle with four rocket stages and a single satellite passenger injected directly into GEO.

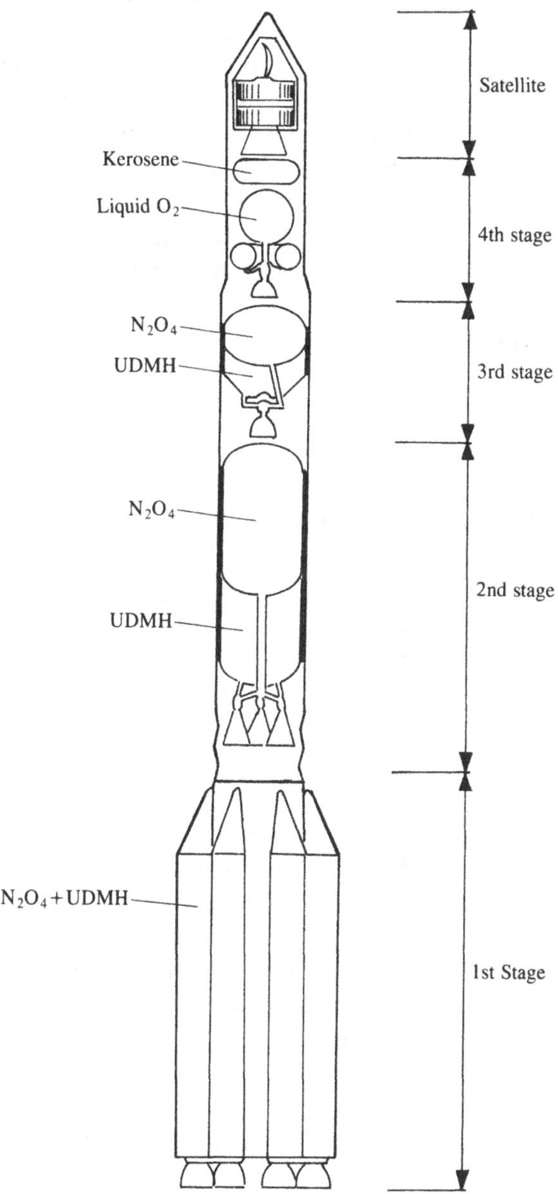

fairly conventional vehicle. The first two stages used UDMH fuel with N_2O_4 as the oxidizer, while the third stage was of the cryogenic H_2/O_2 variety. Ariane-2 was a more powerful version, and Ariane-3 had two solid-fuel strap-on boosters added to the first stage.

Ariane-4 represents a fundamental upgrading of the Ariane-3 concept and is intended as the sole provider of Ariane launch services for geostationary satellites in the 1990s (Fig. 1.15). It is a highly modular vehicle capable of lifting up to three medium-sized satellites into transfer orbit in various tandem arrangements. Liquid and solid propellant strap-on boosters can be configured in six different ways so as to match the thrust to particular satellite constellations.

All versions of Ariane are launched from Kourou in French Guiana, a French possession situated on the northern seaboard of South America. The combination of near-equatorial location, political stability and absence of hurricanes and earthquakes makes French Guiana an ideal launch site for geostationary missions. The climate, which is hot and humid throughout the year, poses no limitations on launch opportunities.

Long March 3 (CZ-3)

In 1949, during the McCarthy era of communist witch-hunting in the US, a young Chinese-American by the name of Tsien Wei Chang was expelled from his research job at the Jet Propulsion Laboratory in California. He decided to leave America and found his way to the newly founded People's Republic of China where he set to work designing missiles. Twenty-one years later, in 1970, China surprised the world by launching a relatively heavy satellite (173 kg) using an all-Chinese rocket – a legacy of Chang's genius and of US anti-communist paranoia.

The CZ-3 vehicle has evolved from earlier versions called CZ-1 and CZ-2. It is the only version proposed by the Chinese to launch foreign geostationary satellites. The CZ-3 resembles Ariane-1 both in appearance and construction (Fig. 1.16). The first two stages are propelled by UDMH/N_2O_4, while the third stage uses cryogenic H_2/O_2.

The flight sequence follows the pattern of Delta-II, except that the CZ-3 sequence is more compressed in time. CZ-3, like the Delta, is suitable only for dedicated launches of small- and medium-sized geostationary satellites.

The vehicle is launched from Xichang in the Sichuan Province some

400 km north-east of the border with Burma. The steep hills and fertile valleys of the region resemble the landscapes of Chinese scroll paintings. Launch opportunities are confined mainly to the winter months due to heavy rainfall during the summer.

Proton SL-12

Proton is the embodiment of brute force. Different versions of this mammoth vehicle have been used for launching lunar missions, orbiting laboratories, geostationary satellites and unmanned low-orbiters. The SL-12 version, which is used for geosynchronous applications, is made up of four stages (Fig. 1.17). The first stage consists of six external tanks attached to a large central tank; each of the external tanks has a rocket motor at the bottom. The second stage is made up of a cluster of four engines fed from two tanks, while the third stage has a single engine. All three stages use UDMH/N_2O_4 propellants. The fourth stage employs a cryogenic mix of kerosene and O_2.

The first three stages place the fourth stage with the satellite onboard in a highly inclined 200-km parking orbit. The fourth stage then performs a PKM function to place the whole assembly in a transfer orbit, and then an AKM function for final injection of the satellite into geostationary orbit. A satellite launched on Proton therefore requires neither a PKM nor an AKM to reach geostationary orbit.

Commercial Proton launches take place from Baikonur in the Kazakhstan Republic 250 miles north-east of the Aral Lake. The launch site is located in a flat, desolate steppe region. The inland climate brings very hot, dry summers and cold winters, neither of which curtail the very intensive launch programme.

Launch Vehicle Reliability

Titan, Atlas/Centaur, Delta and Proton claim launch success rates of over 90%, whereas the figures for Ariane, CZ-3 and STS (inclusive of the upper stage) are somewhat lower. These claims must be taken with a pinch of salt, because the statistical population of comparable vehicle versions is limited. For the old rockets it includes obsolete versions which are drastically different from those currently on offer, while for the newer rockets the total number of launches to date is relatively small. The spate of launch failures in 1986 would also suggest that something has suddenly gone wrong in launch vehicle quality control and project management.

As a conservative rule-of-thumb, one should assume for planning purposes that every fourth or fifth geostationary applications satellite is lost early in life, usually because of a launch failure and, more rarely, due to the satellite's own infant mortality following a successful launch (see Chapter 14). This accident rate is appallingly high, as the satellite insurance community has found. The 1986–1987 launch hiatus was chiefly caused by a growing realization in the space trade that the poor reliability of launch vehicles was upsetting the economic viability of commercial space applications. By the time this book is published, the reliability problem will undoubtedly have been thoroughly analysed, but the cost of remedial action and growing customer insistence on some kind of launch service warranty will bring about a substantial increase in launch prices. The only factors which might curb run-away prices are competition, economy of scale, and inherent limitations in how much prospective satellite owners can afford to pay.

Bibliography

Ariane 4 User's Manual (1983). Issue 1. Evry: Arianespace.

Atlas G/Centaur Mission Planners Guide (1983). San Diego: General Dynamics Convair Division.

Delta Spacecraft Design Restraints (1980). Huntingdon Beach: McDonnell Douglas Astronautics Company.

Long March 3 User's Manual (1985). Beijing: The Ministry of Astronautics.

Vyvedenie Sputnika Radiosvyazi Sovietskoy Raketoy-Nositelem "Proton" (1986). Moscow: Glavkosmos.

2

The Transfer Orbit

Introduction

In this chapter we shall examine the properties of the *geostationary transfer orbit* (GTO). The equations which follow apply to any earth orbit and may in fact be employed to compute low orbits, circular orbits, polar orbits, etc. All parameters are in units of kilometres (km), seconds (s) and radians (rad) unless indicated otherwise.

Kepler's Laws

The German astronomer Johannes Kepler (1571–1630) established three laws governing the movement of planets around the sun. These laws can be applied to the movement of artificial satellites around the earth as follows (Fig. 2.1):

1. Satellites move around the earth in elliptic orbits. The earth is located in one of the focal points of the ellipse, the other focal point is vacant.
2. The radius vector from the earth's centre to the satellite sweeps across equal areas at equal time intervals.
3. The square of the orbital period is proportional to the cube of the semi-major axis.

Thus Kepler's laws define the geometry of orbits, the velocity variation of a satellite along the orbital path, and the time it takes to complete an orbit.

Orbital Geometry

The shape of an arbitrary elliptic orbit is defined by its *semi-*

major axis a and its *eccentricity e* (Fig. 2.2). The geocentric distance of the apogee r_a and the perigee r_p is then obtained as:

$$r_a = a(1 + e), \tag{2.1}$$

Fig. 2.1. Kepler's laws concerning orbital motion.

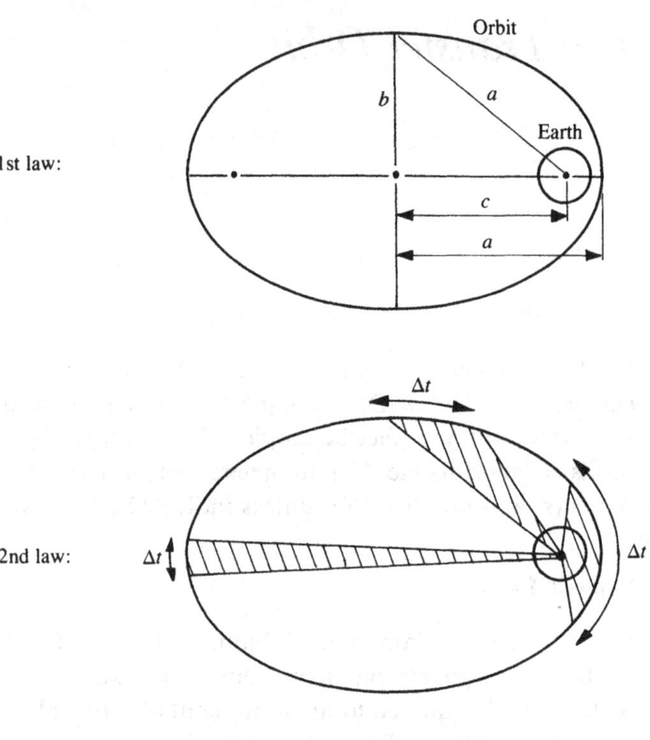

Fig. 2.2. Definition of semi-major axis *a* and eccentricity *e*.

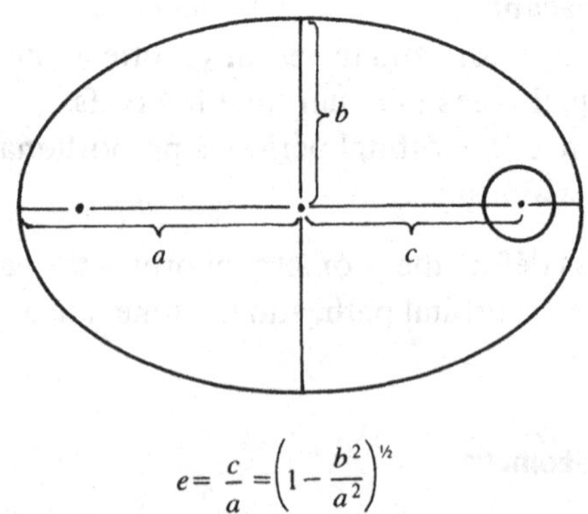

$$e = \frac{c}{a} = \left(1 - \frac{b^2}{a^2}\right)^{1/2}$$

$$r_p = a(1 - e). \tag{2.2}$$

Let R be the earth's radius (R = 6371 km on the average). The apogee height Ap and the perigee height Pe above the earth's surface become:

$$Ap = r_a - R,$$

$$Pe = r_p - R.$$

It follows that

$$a = 0.5(r_a + r_p) = 0.5(Ap + Pe) + R. \tag{2.3}$$

By combining Eqns 2.1 and 2.2, we obtain

$$e = (r_a - r_p)/(r_a + r_p). \tag{2.4}$$

The distance r from the earth's centre to the satellite is:

$$r = \frac{a(1 - e^2)}{1 + e \cos v} \tag{2.5}$$

with v being the *true anomaly* (Fig. 2.3). In Fig. 2.4, r has been plotted against true anomaly for a typical Ariane GTO.

Two other "anomalies" may be used as an alternative to the true anomaly to indicate the satellite's position along the orbit. The *eccentric anomaly E* is defined in Fig. 2.5 and is related to the true anomaly through the relationship:

$$\cos v = \frac{\cos E - e}{1 - e \cos E}. \tag{2.6}$$

The *mean anomaly M* is more difficult to illustrate in a diagram. It is the

Fig. 2.3. Definition of true anomaly v and range distance r.

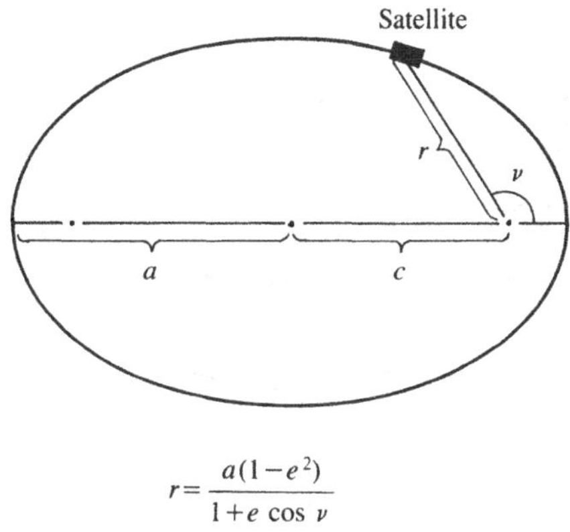

$$r = \frac{a(1-e^2)}{1+e \cos v}$$

Fig. 2.4. Spacecraft range distance r as a function of true anomaly ν and time from perigee for a typical Ariane GTO.

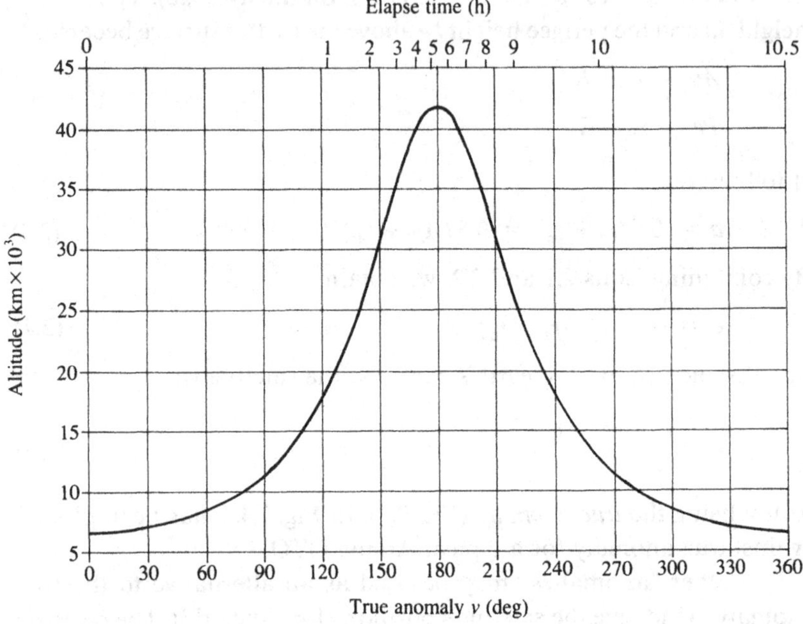

Fig. 2.5. Definition of eccentricity anomaly E.

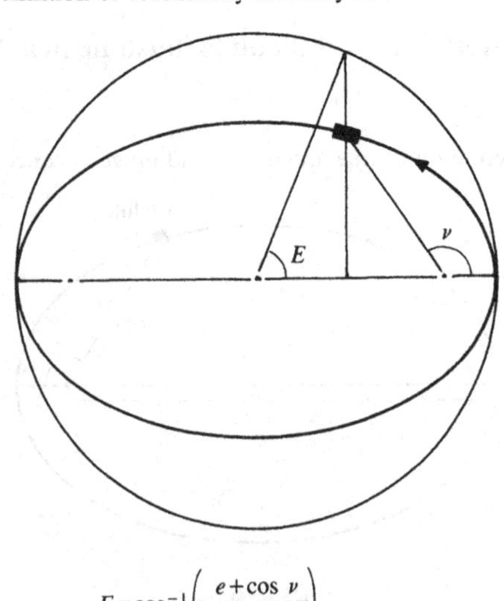

$$E = \cos^{-1}\left(\frac{e + \cos \nu}{1 + e \cos \nu}\right)$$

angular equivalent of the satellite's normalized elapse time t from the last passage through the perigee, such that:

$$M = 2\pi t/\tau \qquad \text{(rad)} \tag{2.7}$$

with τ being the orbital period, i.e. the time it takes for the satellite to complete an orbit. M is also a function of E:

$$M = E - e\sin E \qquad \text{(rad).} \tag{2.8}$$

The relationship between v, E and M for a typical GTO is shown graphically in Fig. 2.6.

Orbital Position in Space

In order to describe fully a satellite's whereabouts in space, we must supplement the orbital geometry with some information about the orbit's orientation. This is best achieved by defining a reference coordinate system which is fixed in inertial space (Fig. 2.7). The x-axis is aligned with the intersecting line between the earth's equatorial plane and the plane of the earth's orbit around the sun (the *ecliptic*). The x-axis is pointing towards the *First Point of Aries*, one of the stars in the constellation Aries. Because the traditional symbol for Aries is a ram's

Fig. 2.6. Eccentric anomaly E and mean anomaly M as a function of true anomaly v and time from perigee for a typical Ariane GTO.

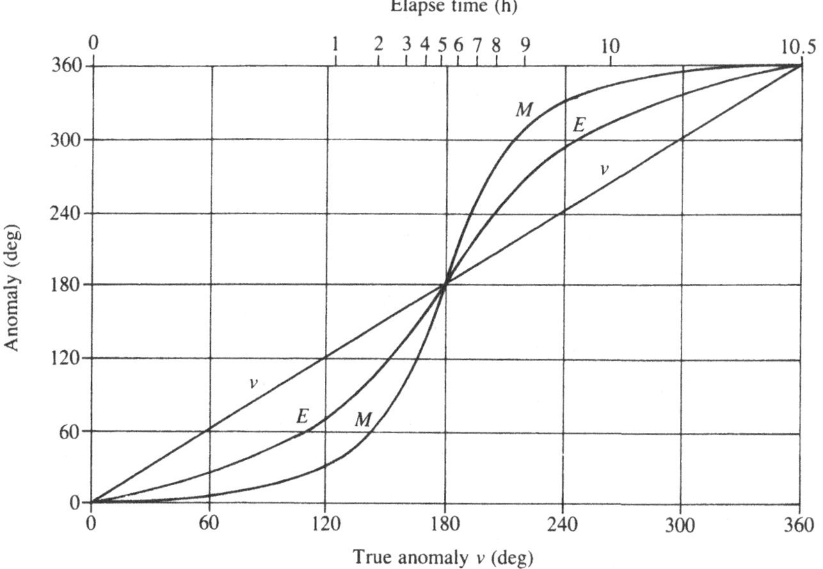

head (♈), this symbol is used in preference to "x". The z-axis coincides with the earth's spin axis and points towards the north. The y-axis completes the right-handed orthogonal coordinate system.

Let us examine Fig. 2.7 further. The *orbital inclination* i is the angle which the orbit forms with the equator, measured on the right-hand side of the *ascending node*. The inclination may thus assume any value from 0 to 180°.

The *right ascension of the ascending node* Ω determines how the orbit is rotated with respect to ♈.

The *argument of perigee* ω is measured from the ascending node and describes how far the orbit is turned within its own plane.

Satellite Position in Space

Given the standard six orbital elements a, e, i, Ω, ω, ν, the position of a satellite in space at any time is uniquely defined. The pair Ap and Pe may be used instead of a and e; similarly, either E or M may replace ν. These six Keplerian elements may seem a strange choice when a set of Cartesian or polar coordinates could have served the same purpose. The main advantage of Keplerian elements is that, if

Fig. 2.7. Definition of right ascension of the ascending node Ω and argument of perigee ω.

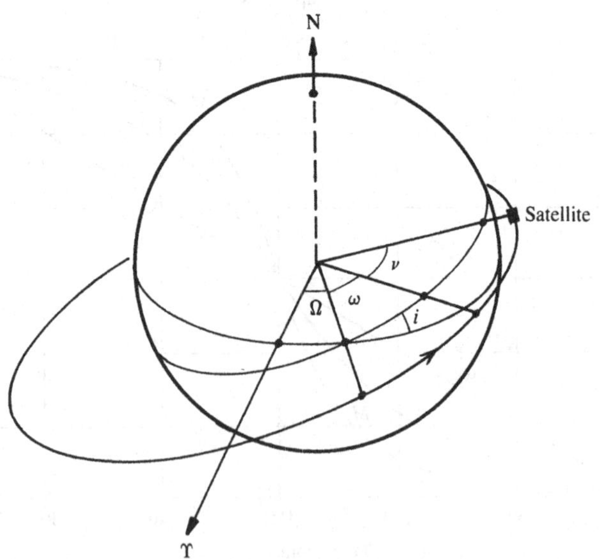

perturbations are disregarded, the first five remain constant with time, while the sixth (if M is selected) varies linearly with time.

Derived Orbital Parameters

The orbital *period* τ is obtained from Kepler's third law as:

$$\tau = 2\pi a \sqrt{a/\mu}. \tag{2.9}$$

Here, μ is the earth's gravitational constant which has a value of 398 601 km^3/s^2. In Fig. 2.8, τ is plotted against a. For a = 42 164 km we have a geosynchronous 24-h orbit. Note that τ is independent of e, i.e. the orbit's eccentricity does not affect the orbital period.

The *velocity* V of the satellite along the orbit is:

$$V = \sqrt{\mu\left(\frac{2}{r} - \frac{1}{a}\right)}. \tag{2.10}$$

A consequence of Kepler's second law is that a satellite travels faster through perigee than through apogee. The variation of V with ν for a typical GTO is shown in Fig. 2.9.

The satellite *flight path angle* η is defined as the angle between the radius vector $\vec{\mathbf{r}}$ and the velocity vector $\vec{\mathbf{V}}$ (Fig. 2.10). We have:

Fig. 2.8. Orbital period τ as a function of semi-major axis a.

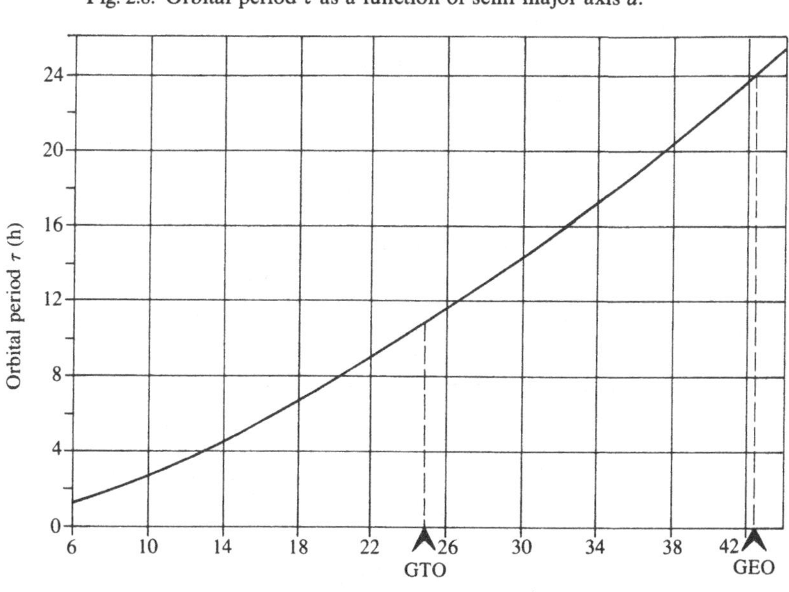

$$\tan \eta = \frac{1 + e \cos \nu}{e \sin \nu}. \tag{2.11}$$

A GTO plot of η against ν is shown in Fig. 2.11.

Fig. 2.9. Orbital velocity V as a function of true anomaly ν and time from perigee for a typical Ariane GTO.

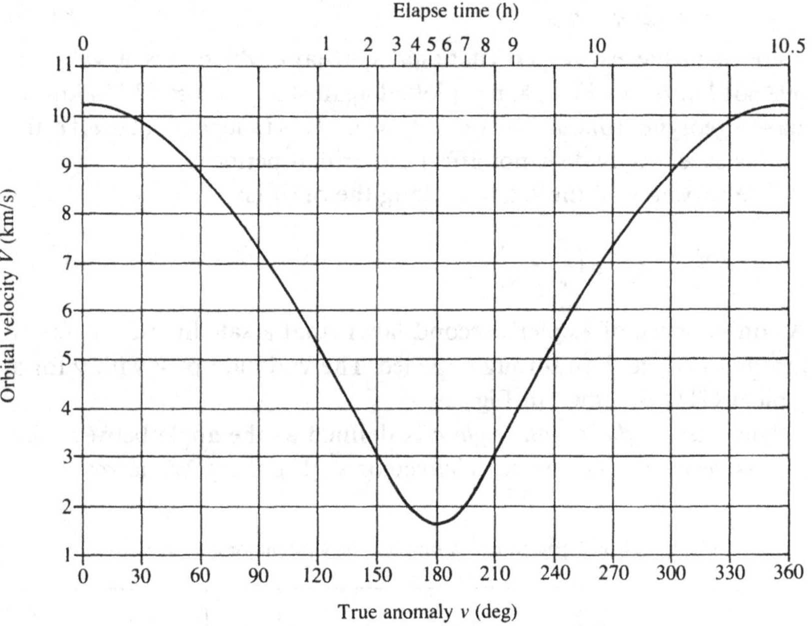

Fig. 2.10. Definition of flight path angle η.

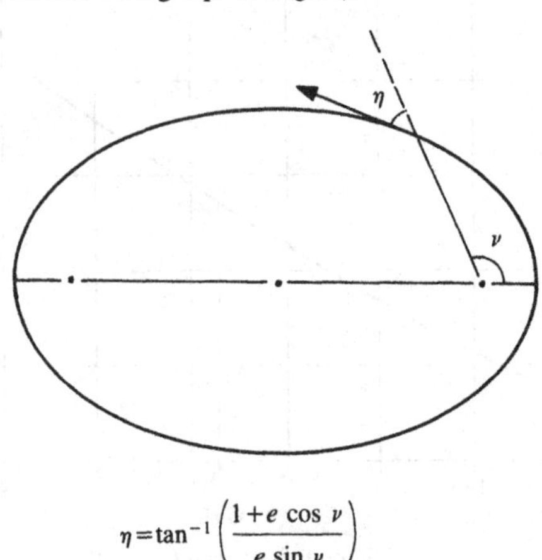

Orbital Perturbations

As we have seen, the elliptic geometry of a Keplerian orbit, and its orientation in inertial space, can be described by five invariant elements a, e, i, Ω, ω. The mathematically "pure" Kepler orbit is, however, a simplified representation of a real satellite orbit. Perturbations in the form of gravitational, geomagnetic and solar pressure forces drive the satellite off course from the Keplerian track. As a result, the real orbit is a somewhat deformed ellipse which is prone to drift with time.

When studying the evolution of an orbit, one may assume that it is perfectly Keplerian at each instant, i.e. that it can be modelled by a unique set of orbital elements which remain constant for a short duration. A modified set is chosen to describe the orbit at the next instant, and so forth. The evolution of a real orbit can thus be modelled using standard orbital elements which vary with time. The time-dependent parameters are called *osculating elements.* Ω and ω are the elements that change most rapidly in a GTO. Let us give these a closer examination.

Fig. 2.11. Flight path angle η as a function of true anomaly ν and time from perigee for a typical Ariane GTO.

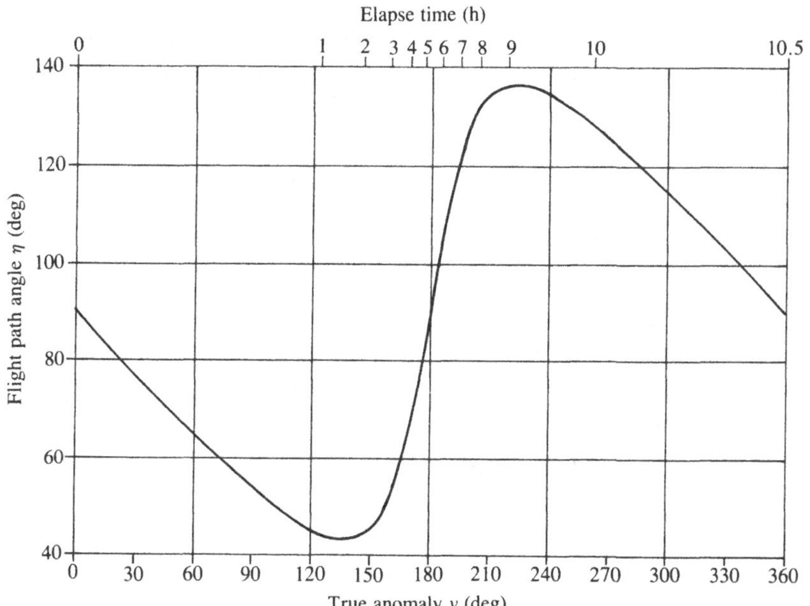

Orbital Precession

The orbit plane possesses gyroscopic properties, i.e. it remains fixed in inertial space unless it is perturbed by external torques. If perturbed, it will begin to drift. The drift is called *orbital precession*. The main perturbing forces are associated with gravitational irregularities.

One very important irregularity is the oblateness of the earth, which causes a precession of Ω at the rate of:

$$\frac{d\Omega}{dt} = \frac{-10}{\left(\frac{a}{r}\right)^{7/2}(1-e^2)^2} \cos i \qquad \text{(deg/day).} \tag{2.12}$$

Note that the precession is *retrograde* (i.e. Ω is decreasing) for $i < 90°$ and *prograde* (Ω is increasing) for $i > 90°$.

Irregularities in the gravitational field also cause the argument of perigee ω to drift:

$$\frac{d\omega}{dt} = \frac{5}{\left(\frac{a}{r}\right)^{7/2}(1-e^2)^2} (5\cos^2 i - 1) \qquad \text{(deg/day).} \tag{2.13}$$

When $(5\cos^2 i - 1) = 0$, or $i = 63.4°$, the drift stops completely, and the apogee and perigee remain stationary. The precession is retrograde when $i > 63.4°$, and prograde otherwise.

Orbital precession is useful in many low or elliptic orbit applications. For example, it allows mission planners to compute *sun-synchronous* orbits where the satellite passes over a given spot on the earth at the same local time every day throughout the year (i.e. Ω is made to precess at the same rate as the earth moves around the sun, or $d\Omega/dt = 0.9856°$ per day). Such orbits are of interest to scientists and earth resources analysts. Another interesting application is the 12-h *Molniya orbit* used by the Russians to keep the apogee fixed (i.e. $d\omega/dt = 0$) above the USSR for maximum ground visibility.

In the special case of a GTO, the magnitude of precession is fairly significant, but the phenomenon brings no operational benefits. For a typical Ariane GTO, with Ap = 35 793 km and Pe = 200 km:

$$d\Omega/dt = -0.416° \text{ per/day,}$$

$$d\omega/dt = 0.821° \text{ per/day.}$$

Sun Angle

During its travels through successive GTOs, a satellite attitude vector adopts a variety of angles with respect to the sun vector, beginning with the attitude at separation from the launch vehicle and ending with the orientation needed to successfully fire the apogee motor. The variation of the *sun angle* θ_s has to be carefully planned and monitored, because the incidence of sunlight onto the satellite affects both the thermal balance of the spacecraft and the amount of electric power available from the solar array.

The angle θ_s is obtained from the scalar product of two unit vectors **z** and **s** (Fig. 2.12), where **z** is the spacecraft attitude vector (typically the spin axis) and **s** is the satellite-sun vector. Because the size of a GTO is very small compared to the distance between the sun and the earth, **s** may be equated to the earth–sun vector. Thus θ_s is obtained from:

$$\cos\theta_s = \mathbf{z} \cdot \mathbf{s}.$$

A fundamental rule in vector algebra is that vectors in a scalar multiplication must be defined in the same coordinate system. We shall adopt a system where the x-axis coincides with the sun–earth vector, the z-axis is aligned with the ecliptic plane normal, and the y-axis completes the right-handed orthogonal coordinate system (Fig. 2.13). Let the orthogonal unit vectors $\mathbf{b}_1, \mathbf{b}_2, \mathbf{b}_3$ represent the axes x, y, z, and call the whole coordinate system B. In this reference frame the unit vector **s** becomes simply:

$$\mathbf{s}_B = (1 \cdot \mathbf{b}_1 + 0 \cdot \mathbf{b}_2 + 0 \cdot \mathbf{b}_3) = (1, 0, 0).$$

The problem is now to find **z** in the same coordinate system B. The most straightforward approach is to start out with a satellite-fixed reference frame I such that (Fig. 2.14):

$$\mathbf{z}_I = (1 \cdot \mathbf{i}_1 + 0 \cdot \mathbf{i}_2 + 0 \cdot \mathbf{i}_3) = (1, 0, 0),$$

and then to perform a series of coordinate transformations so as to fit

Fig. 2.12. Definition of sun angle θ_s and solar aspect angle θ_s'.

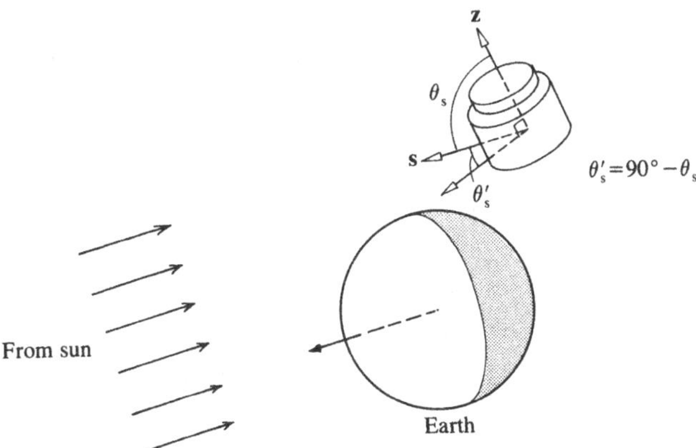

Fig. 2.13. Coordinate system B.

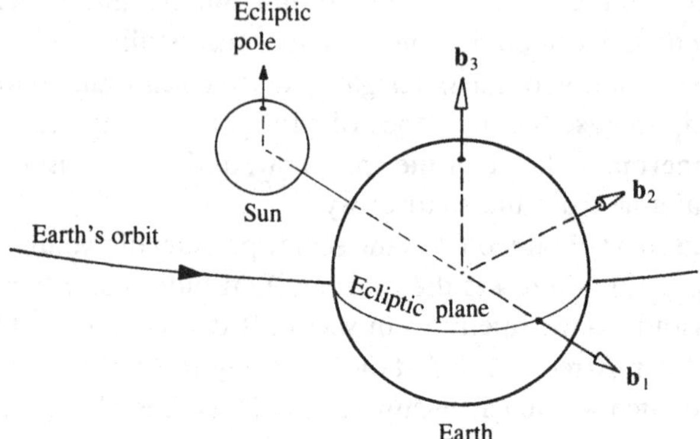

Fig. 2.14. Coordinate system I in relation to system B.

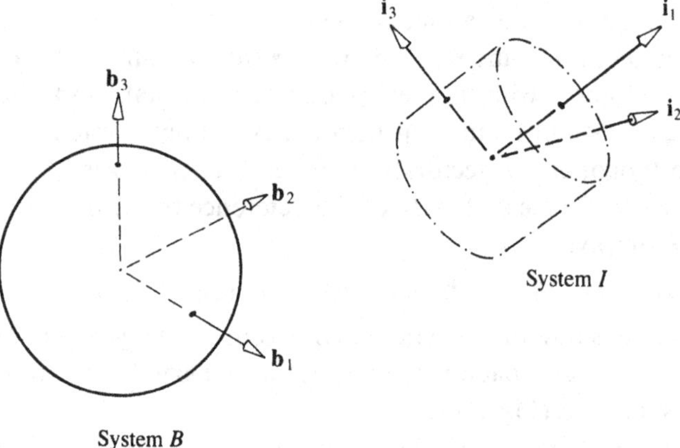

Fig. 2.15. Spacecraft attitude vector azimuth α and elevation δ. $\mathbf{h}_1$ is the projection of $\mathbf{i}_1$ on the orbit plane. $\mathbf{g}_1$ is the unit vector of $\vec{\mathbf{r}}$.

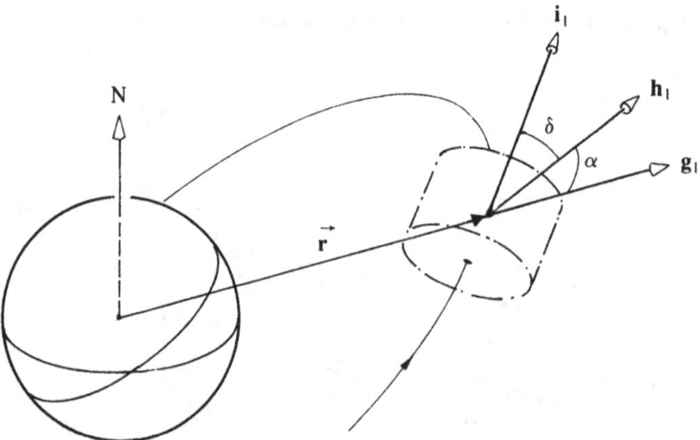

this satellite-fixed frame I into the earth-fixed coordinate system B. This entails rotating the I frame around various axes through a set of angles $\delta, \alpha, \nu, \omega, i, \Omega, \varepsilon, \zeta$. Here δ = spacecraft attitude vector elevation (Fig. 2.15), α = spacecraft attitude vector azimuth (Fig. 2.15), ν = true anomaly, ω = argument of perigee, i = orbital inclination, Ω = right ascension of the ascending node, ε = angle between the earth's equatorial plane and the ecliptic (= 23.44°; Fig. 2.16), and ζ = position of the earth in the ecliptic (Fig. 2.17).

Fig. 2.16. Angle ε between the earth's equatorial plane and the ecliptic.

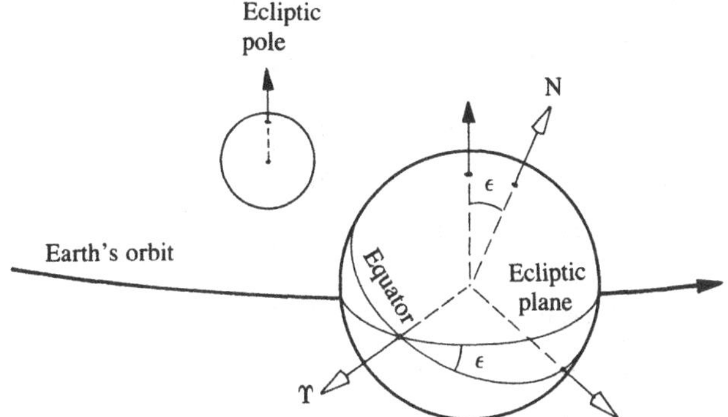

Fig. 2.17. Angle ζ showing the earth's position in its orbit around the sun, measured from the autumn equinox (21 September).

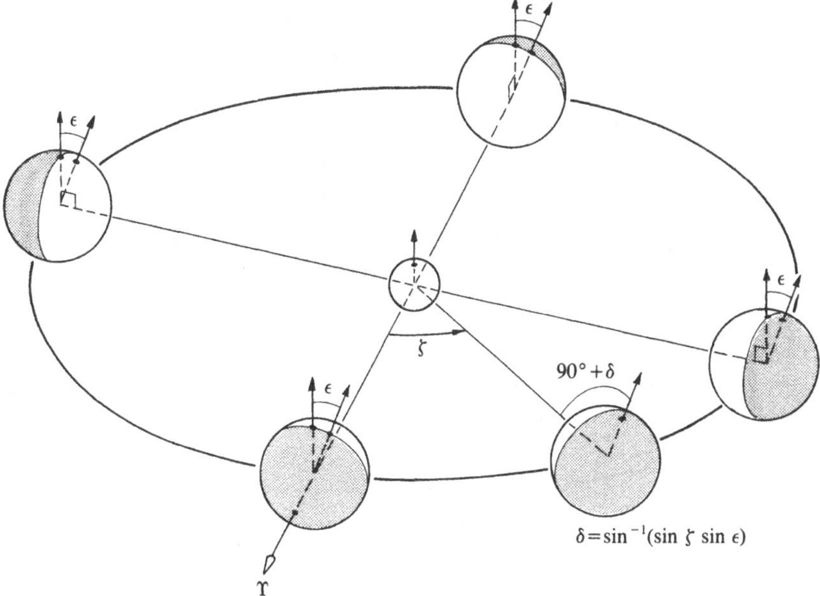

Performing coordinate transformations is a tedious task, and the long trigonometric equations that result are usually of little interest to the reader. We will therefore limit out explanation to an outline of the method used, which is also relevant for later discussions on eclipse transits and sub-satellite paths.

The coordinate system rotational matrices are defined as follows (Fig. 2.18 and 2.19):

$$\mathbf{z}_I = \mathbf{z} = (1, 0, 0):$$

$$I_H = \begin{bmatrix} \cos\delta & 0 & \sin\delta \\ 0 & 1 & 0 \\ -\sin\delta & 0 & \cos\delta \end{bmatrix} H_G,$$

$$H_G = \begin{bmatrix} \cos\alpha & \sin\alpha & 0 \\ -\sin\alpha & \cos\alpha & 0 \\ 0 & 0 & 1 \end{bmatrix} G_F,$$

$$G_F = \begin{bmatrix} \cos\nu & \sin\nu & 0 \\ -\sin\nu & \cos\nu & 0 \\ 0 & 0 & 1 \end{bmatrix} F_E,$$

$$F_E = \begin{bmatrix} \cos\omega & \sin\omega & 0 \\ -\sin\omega & \cos\omega & 0 \\ 0 & 0 & 1 \end{bmatrix} E_D,$$

$$E_D = \begin{bmatrix} 1 & 0 & 0 \\ 0 & \cos i & \sin i \\ 0 & -\sin i & \cos i \end{bmatrix} D_C,$$

Fig. 2.18. Synopsis of coordinate transformations from *I* to *B*.

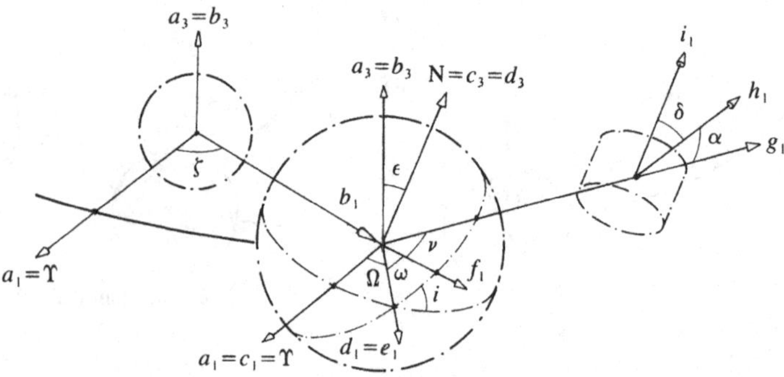

Fig. 2.19. Detailed coordinate transformations from *I* to *B*.

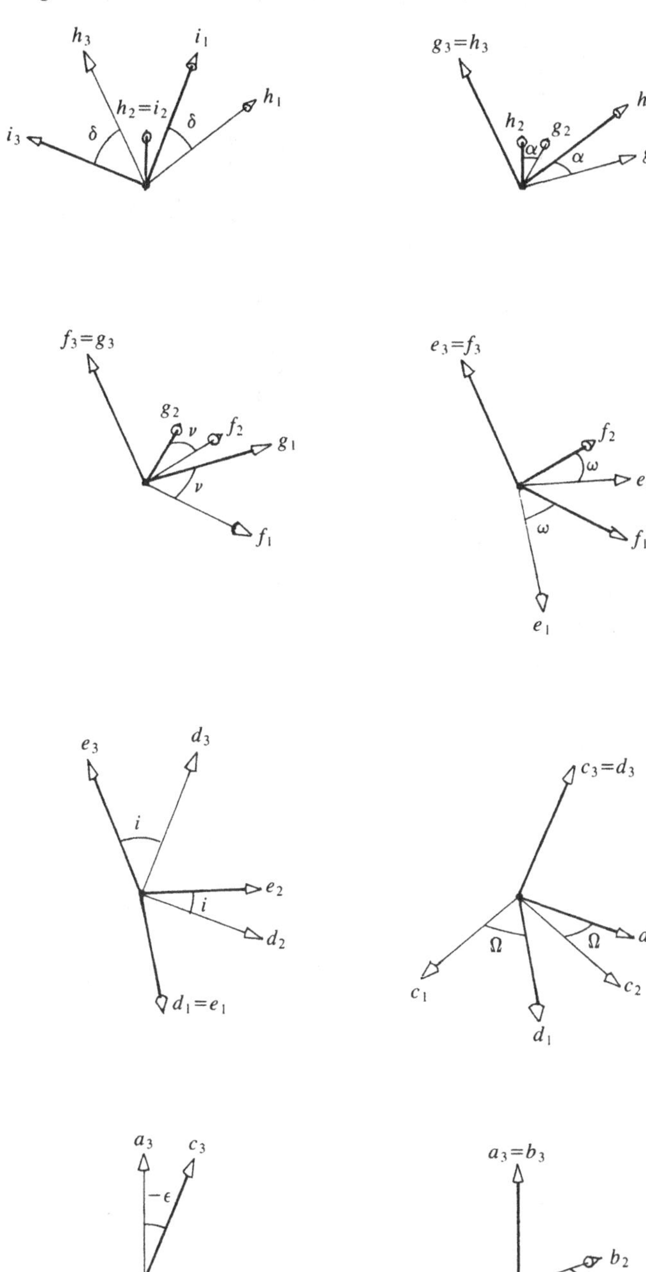

$$D_C = \begin{bmatrix} \cos\Omega & \sin\Omega & 0 \\ -\sin\Omega & \cos\Omega & 0 \\ 0 & 0 & 1 \end{bmatrix} \quad C_A,$$

$$C_A = \begin{bmatrix} 1 & 0 & 0 \\ 0 & \cos\varepsilon & -\sin\varepsilon \\ 0 & \sin\varepsilon & \cos\varepsilon \end{bmatrix} \quad A_B,$$

$$B_A = \begin{bmatrix} \cos\zeta & \sin\zeta & 0 \\ -\sin\zeta & \cos\zeta & 0 \\ 0 & 0 & 1 \end{bmatrix} \quad A,$$

and hence:

$$A_B = B_A^{-1} = \begin{bmatrix} \cos\zeta & -\sin\zeta & 0 \\ \sin\zeta & \cos\zeta & 0 \\ 0 & 0 & 1 \end{bmatrix} \quad B.$$

We may now proceed to bring **z** from the I frame to the B frame by successive matrix multiplications:

Fig. 2.20. Sun angle θ_s as a function of Ω and ζ for a standard Ariane GTO (r_p = 6578 km, r_a = 42 164 km, i = 8°, ω = 180°, δ = 0°, α = 90°).

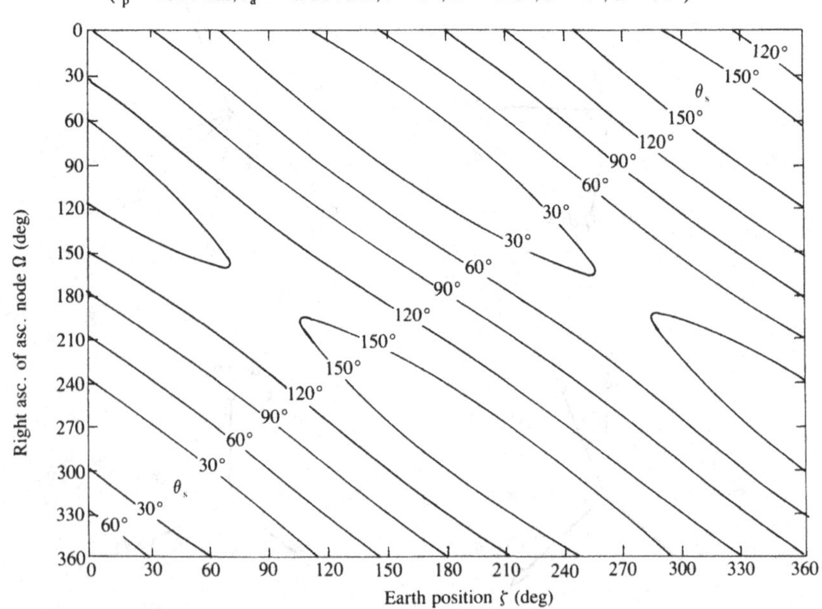

$$\mathbf{z}_B = ((((((((\mathbf{z}_I \cdot I_H) H_G) G_F) F_E) E_D) D_C) C_A) A_B)$$

and finally:

$$\cos\theta_s = \mathbf{z}_B \circ \mathbf{s}_B = \mathbf{z} \cdot \mathbf{s}.$$

The sun angle θ_s is plotted against Ω and ζ in Fig. 2.20 for a typical Ariane transfer orbit. Here, $i = 8°$, $\delta = 0°$, and $\alpha = 90°$ – i.e. the satellite spin axis is tangential to the orbit at apogee in preparation for apogee motor firing (Fig. 2.21).

Eclipse

A satellite is said to be going through eclipse when it passes through the earth's shadow. The shadow is made up of two conical contours known as the *umbra* (total darkness) and the *penumbra* (various degrees of half-shade) (see Fig. 2.22). Seen from the earth, a satellite entering the penumbra gradually loses its white lustre and turns from gold to copper-red before it disappears altogether in the umbra. The copper-red colour is due to the refraction of sunlight as it grazes the earth's atmosphere. A more familiar sight is the copper-red hue of the moon as it passes through the penumbra during a lunar eclipse.

An eclipse creates problems of power supply and thermal control for satellites; moreover, it temporarily prevents satellite sun sensors from performing their attitude determination function. Hence the need to plan and monitor eclipse transits carefully.

Calculating where and when an eclipse occurs requires more steps

Fig. 2.21. Spacecraft attitude at the moment of apogee motor firing.

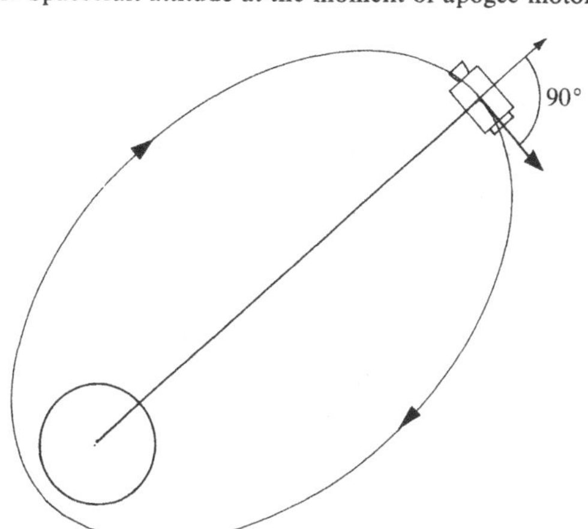

than the computation of sun angle. Basically, the analysis consists in comparing the size and argument of the satellite radius vector $\vec{\mathbf{r}}$ with the conical mantle dimensions of the penumbra and the umbra (Figs 2.23 and 2.24). From Fig. 2.23 it is evident that both cones are symmetrical around the $\mathbf{b}_1$ unit vector, so it would seem logical to repeat the coordinate system transformation process in order to bring the $\mathbf{r}$ vector from the G frame to the B frame. We have:

$$\mathbf{r}_G = (1 \cdot \mathbf{g}_1 + 0 \cdot \mathbf{g}_2 + 0 \cdot \mathbf{g}_3) = (1, 0, 0)$$

Fig. 2.22. Umbra and penumbra geometry.

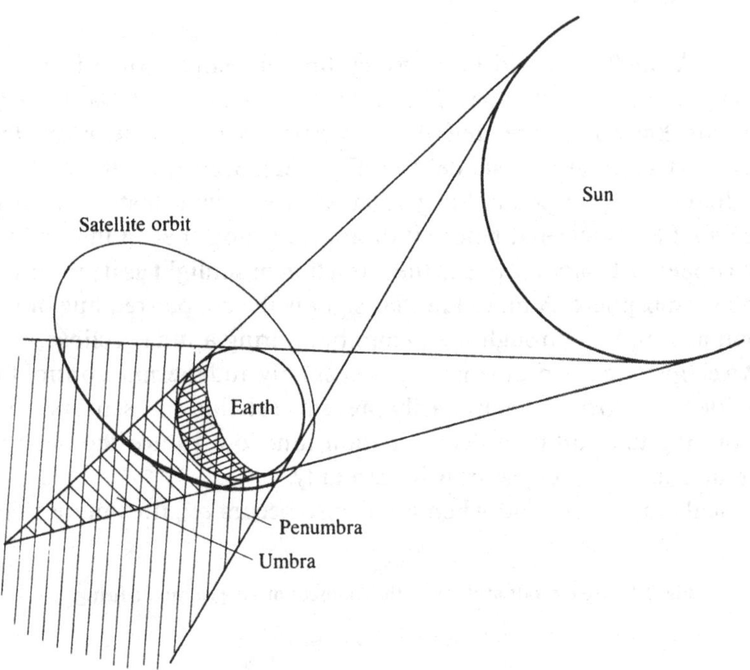

Fig. 2.23. Criteria for penumbra and umbra transits.

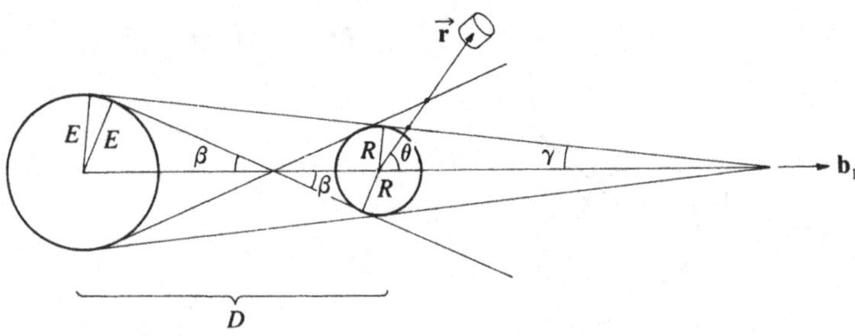

Fig. 2.24. Magnified illustration of eclipse criteria.

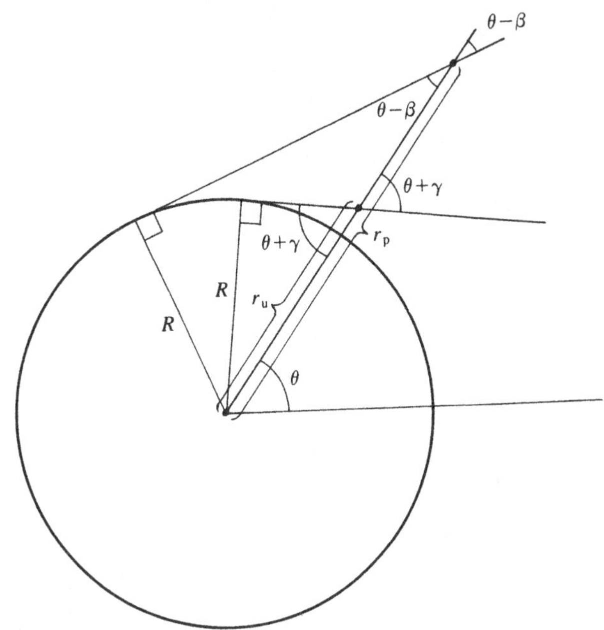

Fig. 2.25. Eclipse duration t_e as a function of Ω and ζ for a standard Ariane GTO.

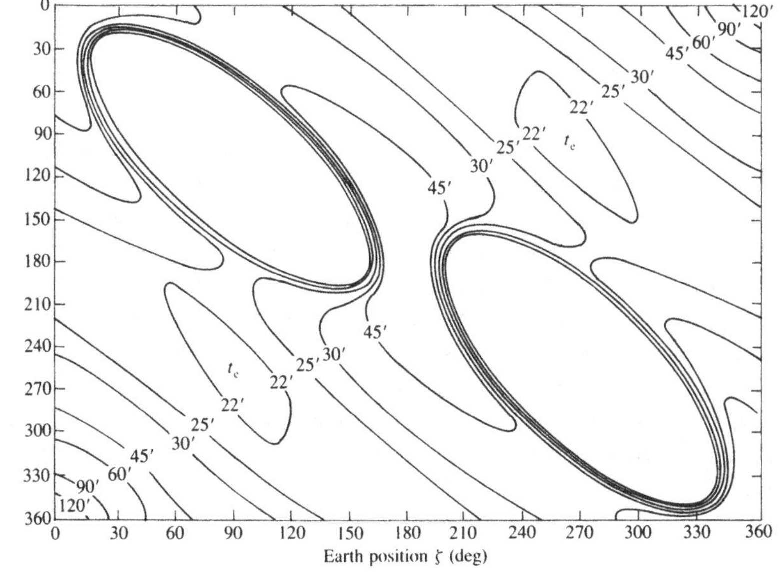

and

$$\mathbf{r}_B = ((((((\mathbf{r}_G \cdot G_F)\, F_E)\, E_D)\, D_C)\, C_A)\, A_B) \tag{2.14}$$

or

$$\mathbf{r}_B = (\mathbf{r}_1, \mathbf{r}_2, \mathbf{r}_3).$$

The conditions for penumbra are (Fig. 2.24):

$$r_u < r < r_p; \qquad \theta < 90° \tag{2.15}$$

and for the umbra:

$$r < r_u; \qquad \theta < 90°. \tag{2.16}$$

Here r is the distance to the satellite as per Eqn 2.5. θ is obtained from:

$$\cos\theta = \mathbf{r}_B \cdot \mathbf{b}_1 = r_1.$$

Finally, to determine the critical range for r, we require:

$$r_u = R/\sin(\theta + \gamma),$$
$$r_p = R/\sin(\theta - \beta)$$

where $\beta = \sin^{-1}\left(\dfrac{E+R}{D}\right) = 0.2688464°$, $\gamma = \sin^{-1}\left(\dfrac{E-R}{D}\right) = 0.2639686°$, $R = 6371$ km (i.e. the average radius of the earth), $E = 695\,893$ km (i.e. the sun's radius) and $D = 149\,597\,871$ km (i.e. the mean distance between the sun and the earth).

Figure 2.25 shows eclipse contours as a function of Ω and ζ for the same Ariane GTO as in Fig. 2.20.

Launch Windows

In order to ensure an adequate power supply and an acceptable thermal environment in GTO, a satellite must be maintained within a specified sun angle and must avoid eclipses longer than a certain duration. For a spin-stabilized satellite in GTO, the sun angle has to be maintained within typically 60–120°. The maximum tolerable eclipse duration could be 22 min. By selecting the 60–120° and 22-min contours from Fig. 2.20 and 2.25, respectively, and by overlaying these, we obtain Fig. 2.26. The overlap areas of acceptable sun angle and eclipse constitute *launch windows*, i.e. the satellite must be launched such that the corresponding values of Ω and ζ are obtained after separation from the launch vehicle.

Sun angle and eclipse are not the only parameters which define launch windows. Depending on the type of satellite and mission, it is often necessary to add constraints such as:

1. True anomaly of eclipse entry and exit (for attitude determination).
2. Colinearity between satellite-earth and satellite-sun vectors (also for attitude determination).
3. Rotation of GTO nodes to achieve certain geosynchronous orbit node values (for inclination drift control).

Fig. 2.26. Launch window in terms of Ω and ζ.

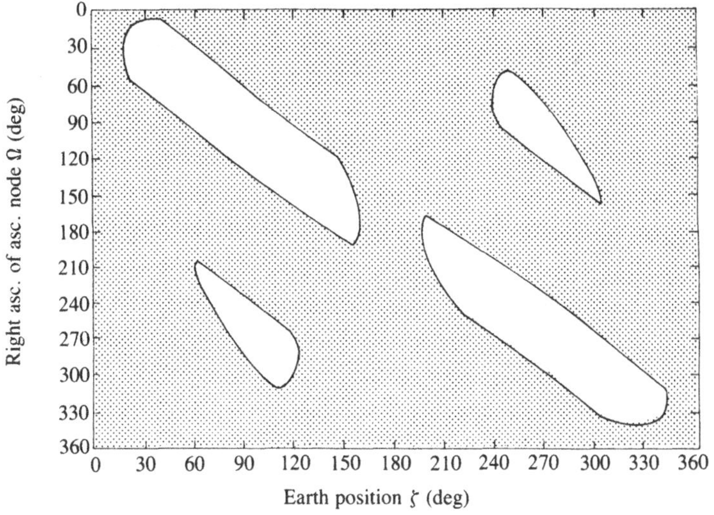

Fig. 2.27. Launch window in terms of launch date *DD* and launch time T_L.

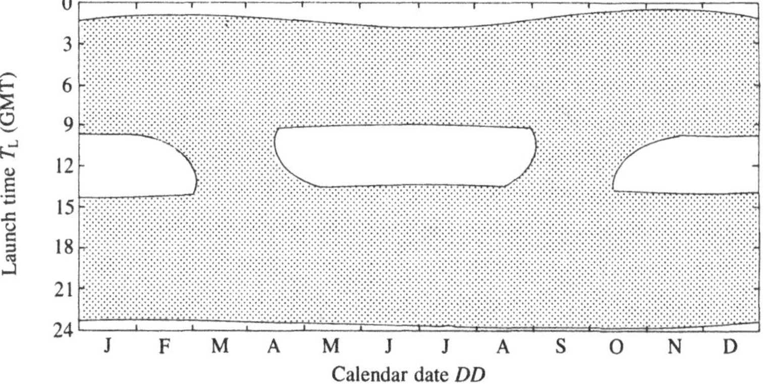

Since sun angles and eclipse durations are partly a function of the position of the sun with respect to the earth – i.e. calendar date and local time – it is customary to put the launch date *DD* instead of ζ along the horizontal axis of the launch window diagram, and launch time T_L (GMT) instead of Ω along the vertical axis (Fig. 2.27). We have:

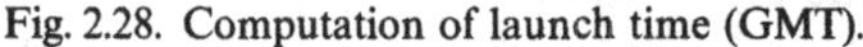

Fig. 2.28. Computation of launch time (GMT).

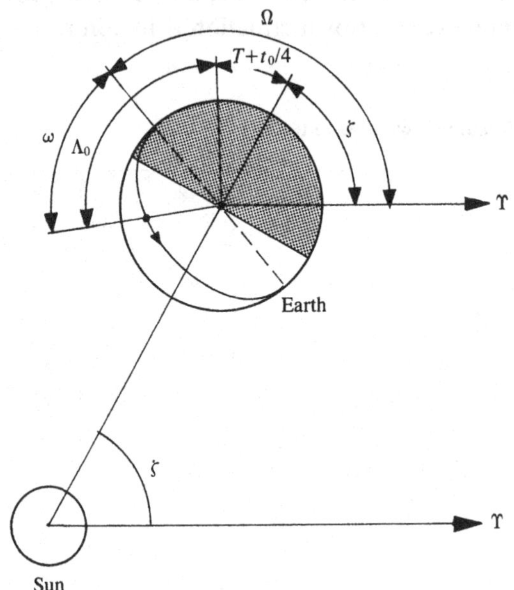

Fig. 2.29. Coordinate system *J*.

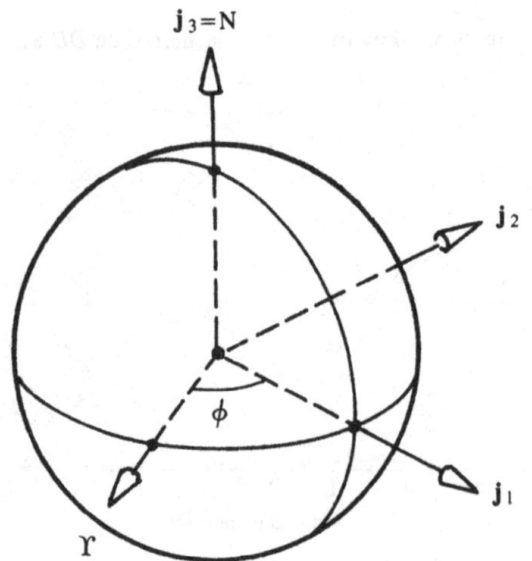

$$\zeta = 360/365.25\,(DD + 101) \qquad \text{(deg)},$$
$$T_L\,(\text{GMT}) \simeq \tfrac{1}{15}(\Omega - \Lambda_0 + \omega - \zeta - t_0/4) \qquad \text{(hours)} \qquad (2.17)$$

where DD = number of days from 1 January (Fig. 2.28), Λ_0 = earth longitude of injection point ($\simeq 0°$ for an Ariane launch), ω = GTO argument of perigee ($\simeq 180°$ for an Ariane launch) and t_0 = time (min) from lift-off until injection ($\simeq 17$ min for Ariane).

Sub-satellite Path

It is useful to draw the projection of the GTO on the earth's surface in order to study the GTO evolution on an earth map. The projection is the trace drawn on the earth by the radius vector as it intersects the surface. The trace is called a *sub-satellite path*. Mission planners use the sub-satellite path to analyse ground station coverage in order to determine telemetry, telecommand and tracking opportunities.

The movements of the earth around its spin axis and of the satellite in its orbit are quite independent. In order to draw the sub-satellite path, it is necessary to relate these movements to each other, using time t as the independent variable. We are therefore once more faced with having to perform coordinate system transformations.

Fig. 2.30. Spacecraft sub-satellite point in terms of direction cosines r_1, r_2, r_3, and of longitude Λ and latitude λ.

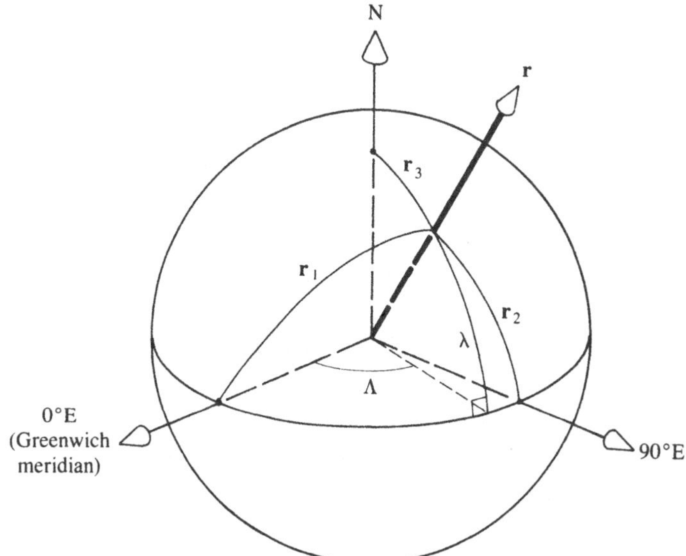

First we need to introduce a new earth-fixed coordinate system J whose x-axis goes through the point on the earth's surface where the Greenwich meridian intercepts the equator; whose z-axis coincides with the earth's spin axis and points north; and whose y-axis completes the right-hand orthogonal coordinate system (Fig. 2.29). This system rotates together with the earth; the x-axis (or $\mathbf{j}_1$ vector) forms an angle ϕ with the intertial vector Υ at any one time. Expressed differently, ϕ is the *sidereal hour angle* of the Greenwich meridian.

We have:

$$J_A = \begin{bmatrix} \cos\phi & \sin\phi & 0 \\ -\sin\phi & \cos\phi & 0 \\ 0 & 0 & 1 \end{bmatrix} A$$

or

$$A_J = \begin{bmatrix} \cos\phi & -\sin\phi & 0 \\ \sin\phi & \cos\phi & 0 \\ 0 & 0 & 1 \end{bmatrix} J.$$

Then:

$$\mathbf{r}_J = ((((((\mathbf{r}_G \cdot G_F)\, F_E)\, E_D)\, D_C)\, C_A)\, A_J)$$

or

$$\mathbf{r}_J = (r_1, r_2, r_3).$$

The sub-satellite path, in terms of latitude λ and longitude Λ, becomes (Fig. 2.30):

$$\lambda = \sin^{-1}(r_3),$$

Fig. 2.31. Sub-satellite path of a typical Ariane early GTO.

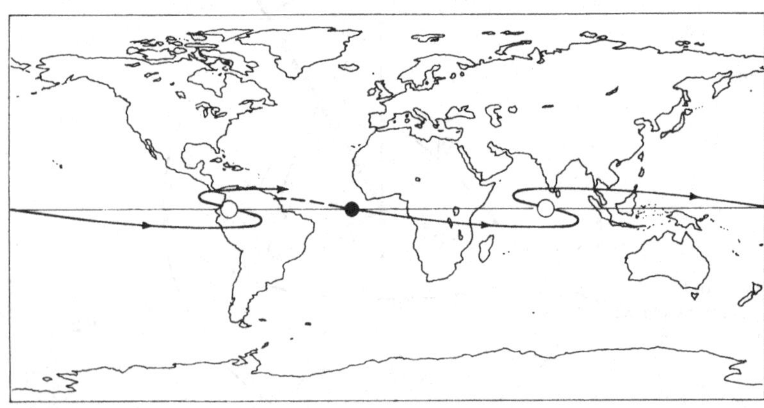

●=perigee; ○=apogee.

$$\Lambda = \tan^{-1}(r_2/r_1).$$

A typical GTO sub-satellite path for an Ariane launch is shown in Fig. 2.31.

Bibliography

Bohrmann, A. (1966). *Bahnen Künstlicher Satelliten*, Hochschultaschenbücher 40/40a. Mannheim: Bibliographisches Institut.

Giese, R. H. (1966). *Weltraumforschung I*, Hochschultaschenbücher 107/107a. Mannheim: Bibliographisches Institut.

Le mouvement du satellite (1984). Conférences et exercices de mécanique spatiale 1983. Toulouse: Cepadues-Editions.

3

The Geostationary Orbit

Introduction

A *geosynchronous orbit* is any earth orbit with an orbital period equal to the time it takes the earth to rotate once on its axis, i.e. 23 h and 56 min. We recall from Eqn 2.8 that the period is a function only of the semi-major axis a, and is independent of the other five standard orbital elements. A geosynchronous orbit may thus be elliptic or inclined to an arbitrary degree, as long as a = 42 164 km. In the short term, the sub-satellite path describes a more or less distorted figure "8" centred around a fixed point on the equator (Fig. 3.1); in the longer term, the point does not remain fixed but drifts along the equator.

Fig. 3.1. Sub-satellite path of a circular, inclined GEO. For small i: $\Delta\Lambda = i^2/4$ and $\lambda_{\Delta\Lambda} = i/\sqrt{2}$.

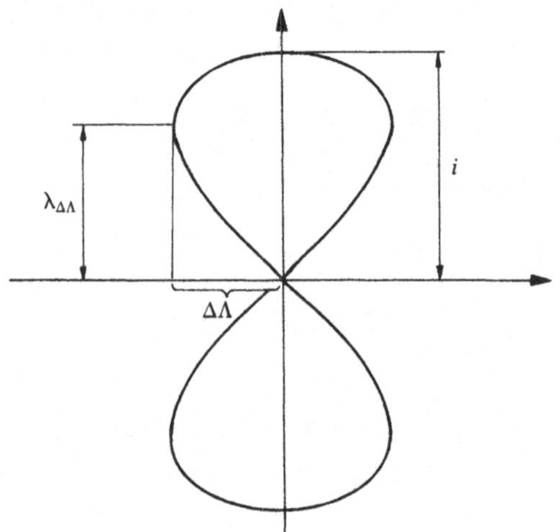

A *geostationary orbit* is a special case of the geosynchronous variety with i and $e = 0$. A truly geostationary orbit is therefore perfectly circular and equatorial. Seen from the ground, a geostationary satellite remains fixed at the same point in the sky day and night. The sub-satellite path is reduced to a point on the equator.

The adjectives "geosynchronous" and "geostationary" are often used interchangeably by people in the trade. We shall not belabour these definitions, but "geostationary" will be used consistently in the following to denote near-circular orbits with inclinations of less than 5°.

A geostationary orbit (GEO) is a special case of the generalized elliptic orbit described mathematically in Chapter 2. The same equations are therefore equally valid for GEO. So why bother with a separate Chapter to describe it? The answer is that most of the equations can be greatly simplified, and certain perturbations are of interest in that they create specific operational problems.

Orbital Geometry and Position in Space

In a circular orbit, the eccentricity e is equal to zero, and therefore the concepts of apogee and perigee have no meaning. According to Eqn 2.5, the radius r is constant and equals the semi-major axis of a. From Eqns 2.6 and 2.8 it is also evident that v, E and M are identical and are a simple function of orbital time t. The argument of perigee ω is undefined since there is no perigee. As we shall see, the right ascension of the ascending node Ω becomes undefined when the inclination i approaches zero.

Derived Orbital Parameters

It may seem surprising that the orbital period we desire from a geostationary satellite is not 24 h, but 23 h and 56 min. As the earth rotates around the sun while spinning around its own axis, the period of one turn of the earth is different with regard to the sun (24 h) than to the stars (23 h and 56 min) (see Fig. 3.2). Since orbits are defined in inertial space, i.e. in the "star" frame, the period of a truly geostationary orbit coincides with the *sidereal* ("star") day rather than the *calendar* ("sun") day. Stated differently, the calendar day is the 24 h it takes the earth to rotate 360° with respect to the earth–sun vector, whereas the earth's sidereal day is the 23 h and 56 min required for a complete rotation of the earth in relation to the inertial Υ-vector.

From Eqn 2.9:

$$a^3 = \mu\,(\tau/2\pi)^2$$

With τ = 23 h and 56 min or, more precisely, τ = 86 164 s, we obtain:

$$a = 42\,164 \text{ km}.$$

Equation 2.10 for satellite velocity is simplified to:

$$V = \sqrt{\mu/a} = 3.075 \text{ km/s}.$$

Fig. 3.2. Calendar day and sidereal day. 1 calendar day $\triangleq$ 361° $\triangleq$ 24 h and 1 sidereal day $\triangleq$ 360° $\triangleq \dfrac{360}{361} \times 24$ h = 23 h 56 min.

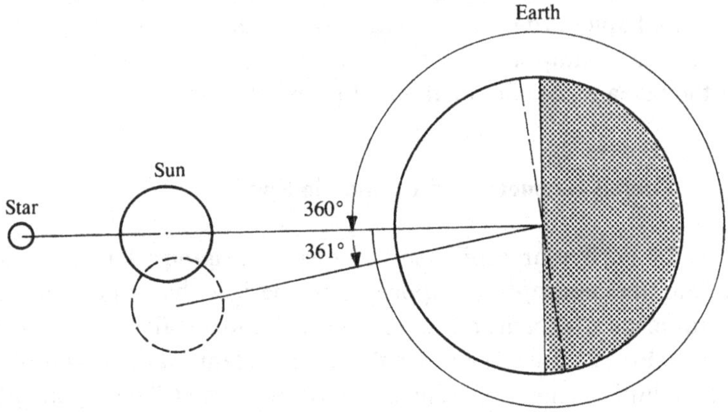

Fig. 3.3. Cross-section of the earth's shadow at GEO altitude.

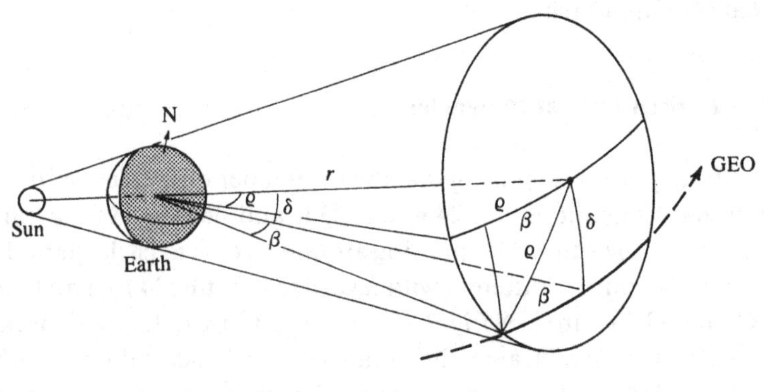

$\epsilon = 23.44°$

$\varrho = 8.70°$

$\delta = \sin^{-1}(\sin \zeta \sin \epsilon)$ — see also Fig. 2.17

$\beta = \cos^{-1}(\cos \varrho / \cos \delta)$

$t_e = 8\beta$ (minutes)

The flight path angle η is 90° throughout the circular orbit, as can be verified by setting $e = 0$ in Eqn 2.11.

The precession of ω is of no interest since there is no perigee. The precession of Ω and the variations of i and a with time do interest us, however, as we shall see later.

Eclipse

While a satellite in GTO almost invariably finds itself in eclipse somewhere along the orbital path, the encounters in GEO are limited to a six-week period centred around the vernal and autumn equinoxes.

The eclipse duration t_e is easy to calculate for the circular GEO. If no distinction is made between the umbra and the penumbra (the latter is relatively brief except during the first and last few days of an eclipse season), the earth's shadow may be represented by a cylindrical contour, and the computation of eclipse duration then becomes even simpler.

Assume the inclination to be $i = 0°$. Let ρ be the angular radius of the shadow (Fig. 3.3). If R is the earth's radius at the equator (= 6378 km) and r is the geocentric distance to the satellite (= 42 164 km), then:

$$\rho = \sin^{-1}(R/r) = 8.70°.$$

Now let ζ represent the earth's position in its orbit around the ecliptic, measured from the point where the orbit intersects the ♈-vector. If DD is the number of days from 1 January, then:

$$\zeta = (DD + 101)\,360/365.25 \qquad \text{(deg)}$$

Inspection of Fig. 3.3 gives:

$$\delta = \sin^{-1}(\sin\zeta \sin\varepsilon) \qquad \text{(deg)}; \quad \varepsilon = 23.44°,$$

$$\beta = \cos^{-1}(\cos\rho/\cos\delta) \qquad \text{(deg)}$$

and

$$\tfrac{1}{2}t_e = \beta\tau/360 = \beta\,1440/360 = 4\beta \qquad \text{(min)}.$$

Therefore

$$t_e = 8\beta \qquad \text{(min)}.$$

For example, if $\zeta = 0°$ or $180°$, then $t_e = 72$ min, which constitutes the maximum eclipse duration at vernal and autumn equinox. The eclipse season begins and ends with $\beta = 0°$. Then $\rho = \delta$, and $\zeta = 22°$. Since $1° \simeq 1$ day, we find that the eclipse season lasts ± 22 days around 23 September and 21 March (Fig. 3.4).

North–South Drift

As mentioned earlier, the sub-satellite path of a geostationary satellite in a slightly inclined orbit describes a figure "8" around the nominal longitude on the equator (Fig. 3.1). The half-amplitude of the figure equals the orbital inclination. The satellite traverses the loop once every sidereal day. This loop movement is undesirable as it complicates satellite tracking, induces Doppler effects on radio link frequencies, and may result in loss of contact with the ground stations at the limits of visibility.

A simple solution would be to bring the inclination to zero and keep it there. The inclination tends to drift, however, a phenomenon known as *north–south* drift. Maintaining a zero inclination entails constant orbit manoeuvring, spending precious satellite propellant in the process. In space applications where a certain inclination can be tolerated, satellites tend to be deliberately injected into inclined orbits such that i drifts downwards towards zero initially before turning around and moving up again. There is thus an *inclination window*, inside which the inclination drifts freely, thereby obviating the need for inclination control, or *north–south station-keeping*, for up to several years. Once the orbit begins to leave the inclination window, it is necessary to give it a boost back in again (see Chapter 9).

The main cause of the secular drift of orbital inclination is the gravitational pull of the sun and the moon on the satellite. The effect of the gravitational forces varies depending on the orientation of the orbit,

Fig. 3.4. Annual eclipse profile.

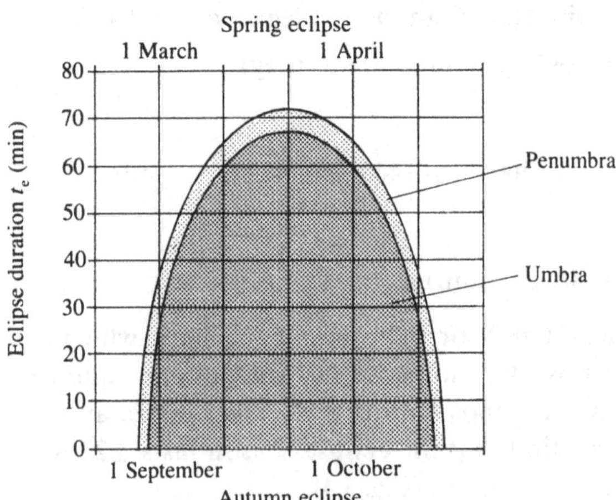

i.e. on Ω. When we talk about inclination drift, we are therefore also interested in how it relates to the precession of Ω.

Figure 3.5 illustrates the relationship between i and Ω for some drift cases. We immediately discover two interesting effects:

(a) The inclination *decreases* initially if we start off with Ω in the range 180–360°, and *increases* for other initial values of Ω. The case of decreasing inclination is of primary operational interest. As the orbit approaches minimum inclination, Ω begins to swing around rapidly towards a value which is up to 180° away from the initial position; from then onwards the inclination will continue to decrease.

(b) The inclination does not necessarily go through zero in the process of drifting from one end of the inclination window to the other.

The slight periodic wobble on each trace in the polar diagram is caused by the attraction of the sun. Since the wobble period is 6 months, it provides a convenient method of measuring time along a trace.

Fig. 3.5. Inclination vector traces on the i–Ω diagram.

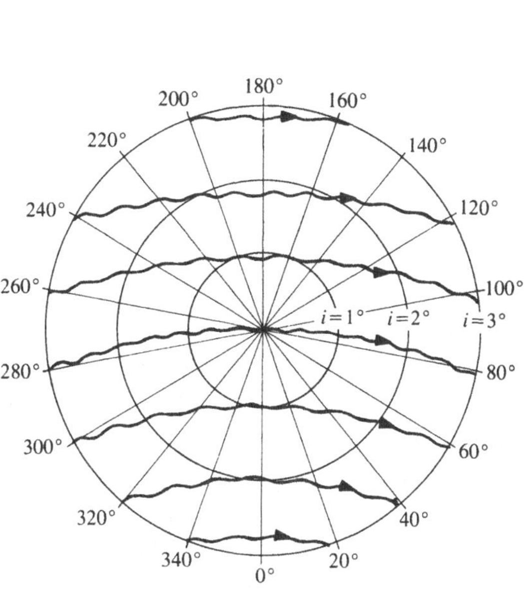

Inclination and nodal drift is best illustrated by introducing the concept of *inclination vector.* This vector is simply the orbit plane normal, as defined in an earth-centred inertial coordinate system, whose x-axis points to the First Point of Aries, whose z-axis points north, and whose y-axis completes the right-handed orthogonal system. Thus:

$$\mathbf{i} = (\sin i \cos \Omega, \sin i \sin \Omega, \cos i)$$

In order to obtain a better physical understanding of the inclination vector's movement through the i–Ω diagram, try to imagine concentric circles of constant i drawn around the north pole of a celestial sphere surrounding the earth. The Ω "spokes" correspond to longitudes on the sphere, counting eastward from ϒ. The orbit plane normal vector then draws traces on the sphere as i and Ω begin to drift (Fig. 3.6).

A reasonably accurate algorithm for computing the drift of i and Ω may be developed from the knowledge that the i vector describes a cone around a vector tilted 7.4° away from the earth's spin axis towards the ecliptic pole, and that it takes 54 years for the vector to complete a cycle (Fig. 3.7). Place a simple x/y coordinate system on the flattened celestial sphere around the North Pole. Let x and y be the projections of the inclination vector on the x- and y-axis, respectively, such that $x = i \cos \Omega$ and $y = i \sin \Omega$. Assume that i_0 and Ω_0 are initial values at the beginning of life. Then:

$$\frac{d\phi}{dt} = \dot{\phi} = 2\pi/54 \text{ (rad/year)} = \pi/324 \text{ (rad/month)},$$

Fig. 3.6. Inclination vector trace on the celestial sphere.

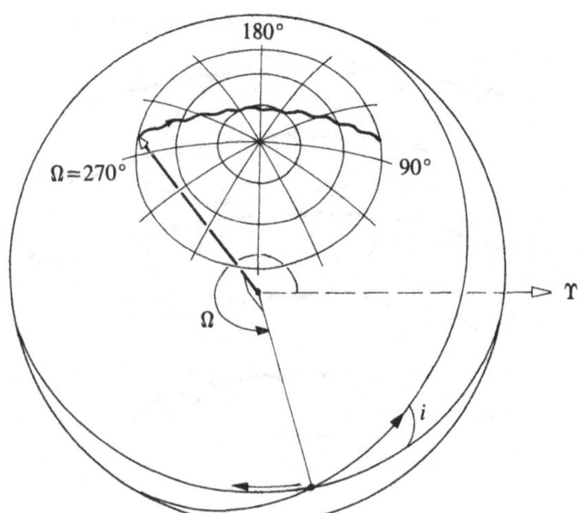

$$\begin{aligned}
\phi &= \phi_0 + \dot{\phi} t \\
\phi_0 &= \tan^{-1}\left(\frac{i_0 \sin \Omega_0}{\theta - i_0 \cos \Omega_0}\right); \qquad \theta = 7.4°, \\
\rho &= \frac{i_0 \sin \Omega_0}{\sin \phi_0}, \\
\Omega &= \tan^{-1}\left(\frac{\rho \sin \phi}{\theta - \rho \cos \phi}\right) \\
i &= \frac{\rho \sin \phi}{\sin \Omega}
\end{aligned} \tag{3.1}$$

The rate of inclination drift $\mathrm{d}i/\mathrm{d}t$ near $i = 0°$ varies from year to year according to:

$$\frac{\mathrm{d}i}{\mathrm{d}t} = \sqrt{u^2 + v^2} \qquad \text{(deg/year)} \tag{3.2}$$

where $u = -0.1314 \sin \Omega_m$ (deg/year), $v = 0.8541 + 0.09855 \cos \Omega_m$ (deg/year) and $\Omega_m = 12.111 - 0.052954\, t$ (deg).

Ω_m is the right ascension of the moon's ascending node (period = 18.6

Fig. 3.7. Long-term movement of the inclination vector.

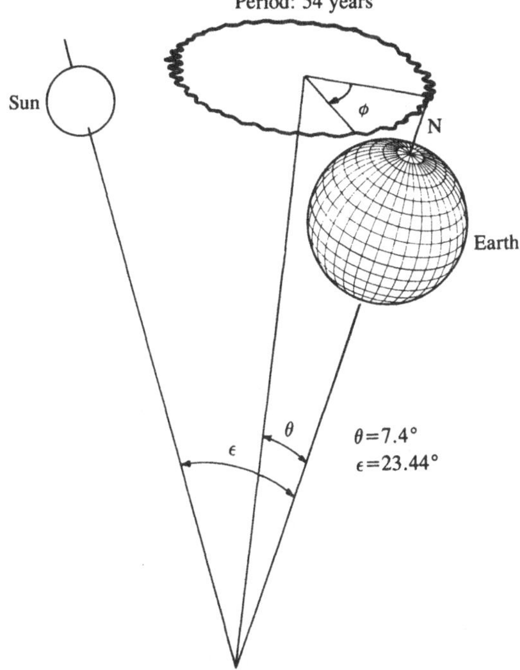

years). Equation 3.2 puts di/dt in the interval between 0.756 and 0.952°/year.

East–West Drift

Equation 2.9 demonstrated that the orbital period is a function of the semi-major axis a. Because the earth's gravity varies slightly with latitude λ and longitude Λ, a satellite experiences gravity gradients at various points above the earth. The gradients give rise to accelerations or decelerations as if the satellite were rolling around on the bumpy surface of the earth's gravity potential function.

Rather than carrying out the somewhat lengthy differentiation of the potential function, let us accept that a very crude mathematical model of east–west acceleration of a geostationary satellite along the equator is given by:

$$\ddot{\Lambda} = C \sin (150° - 2\Lambda) \qquad (\text{deg/day}^2) \tag{3.3}$$

where

$$C = \{16.9 + 2.9 \sin (\Lambda - 35°)\} \times 10^{-4}.$$

Equation 3.3 is shown graphically in Fig. 3.8. Evidently there are four points on the equator where the satellite experiences zero perturbing

Fig. 3.8. Satellite east–west drift acceleration as a function of longitude.

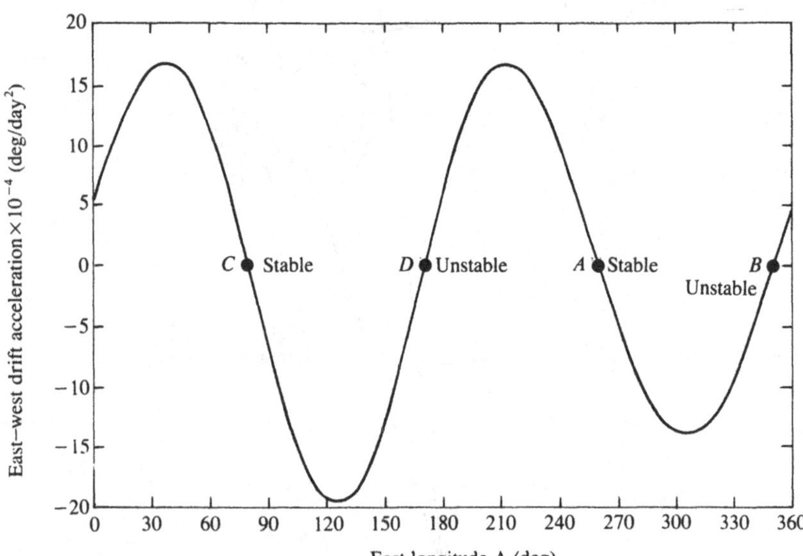

acceleration (points *A, B, C* and *D* in Fig. 3.8, or at approximately 105°W, 15°W, 75°E and 165°E in Fig. 3.9). A satellite positioned over any of these four points will not move, whereas at all intermediate longitudes it will accelerate towards the east or the west. An important consequence is that regular *east–west station-keeping* will be necessary if the satellite is to be maintained within a band centred around an intermediate longitude (the *longitude window*). As in the case of north–south station-

Fig. 3.9. Satellite east–west drift directions. *A* and *C* are stable points, while *B* and *D* are unstable.

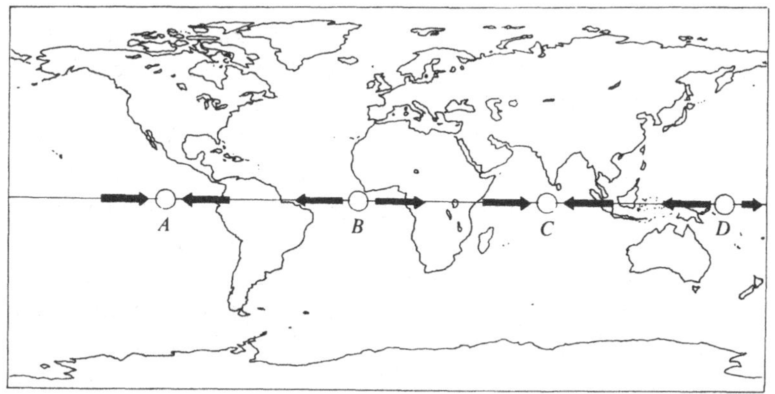

Fig. 3.10. Satellite east–west drift velocity and period as a function of longitudinal distance from stable point.

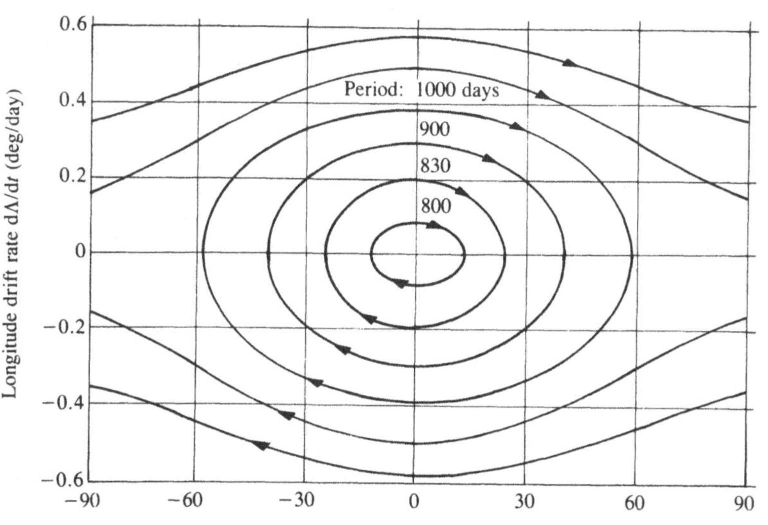

keeping, thrusters will have to be activated and propellants consumed, although east–west corrections consume far less propellant than north–south manoeuvres.

If we now move along the longitude axis in the positive sense in Fig. 3.8, we discover another interesting fact. As we approach points *A* or *C*, the acceleration increases at an ever slower rate. Once we have passed the point in question, the acceleration turns into a deceleration at a growing rate. This means that points *A* and *C* are *stable*, i.e. if a satellite located at either of these two points is displaced by external forces (such as luni-solar gravity), it will merely oscillate around the stable point linke a pendulum.

By similar reasoning we find that points *B* and *D* are *unstable*, so that a perturbed satellite will gradually slide away from an unstable point and begin to oscillate around the nearest stable point (i.e. *A* or *C*).

The mid-points between the zero drift inflexions in Fig. 3.8 represent points of maximum acceleration. A satellite located in any of these regions would have to expend more propellant more frequently than at any other location to remain within the specified longitude window.

Knowing the angular velocity $\dot{\lambda}$ of a satellite as a result of east–west drift is of interest in order to calculate ΔV, i.e. the orbital impulse and hence the propellant mass needed to push the satellite back inside the specified longitude window (usually $\pm 0.1°$ around nominal). Figure 3.10 shows the variation of $\dot{\lambda}$ with longitudinal distance from a stable point for a freely oscillating satellite. The cyclic period is also indicated.

In the real world, the stable points are in fact located at 104°W and 74°E, while the unstable points are at 12°W and 162°E. The method of calculating ΔV is explained in Chapter 9.

Bibliography

Soop, E. M. (1983). *Introduction to Geostationary Orbits*, ESA SP-1053. Paris: European Space Agency.

4

The Satellite Environment

Introduction

From launch onwards, the quality of life of a satellite is abysmal. During the ascent phase it is subjected to violent acceleration, vibration, shock and decompression which stretch its endurance to the limit – and that is only the beginning of a satellite's troubles.

On earth, vacuum is employed to extend the storage life of foodstuffs. Out in space, vacuum has the opposite effect on satellites, for it shortens their lifespan. In the absence of an atmosphere, they are bombarded with charged particles and exposed to ultraviolet radiation. Different parts of a satellite reach temperature extremes at the same time and, because there is no temperature exchange through convection, such extremes cause structural stress leading to possible malfunction. The particle bombardment gives rise to electrostatic discharge which produces short or open circuits and burns out electronic components. Lubricants evaporate in vacuum and cause moving parts to seize up. Paints and sealants "outgas" (perspire) and settle on sensitive optical surfaces. Micrometeorites travel unimpeded through space and strike satellites with tremendous impact.

Fortunately, a satellite's environment is largely predictable. Much of the time and money spent on building a spacecraft goes on verifying its resilience against a known environment through elaborate quality control and testing. In this chapter we shall explore the environment surrounding a geostationary satellite during all its phases of flight.

Powered Flight Loads

During the ascent phase, a satellite is subject to compression forces due to *quasi-static acceleration*. It is measured in "g", i.e. as a scale

Fig. 4.1. Long March 3 acceleration during the ascent phase.

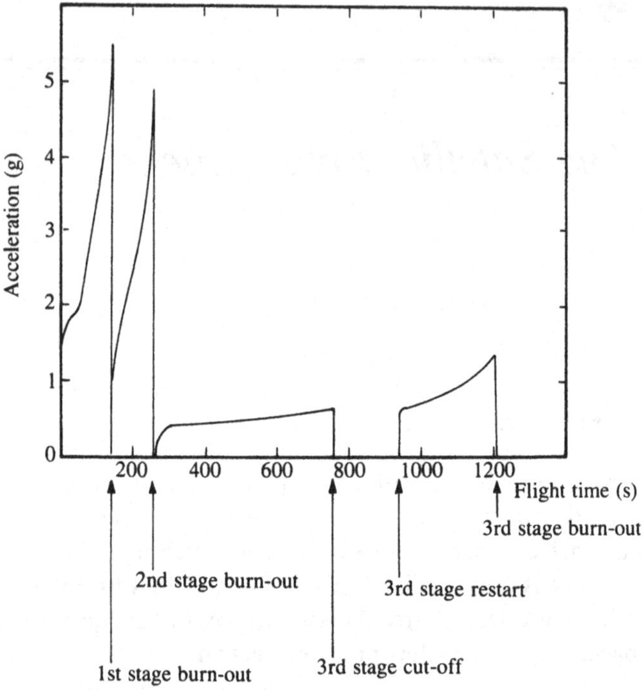

Fig. 4.2. Ariane-4 acceleration during the ascent phase.

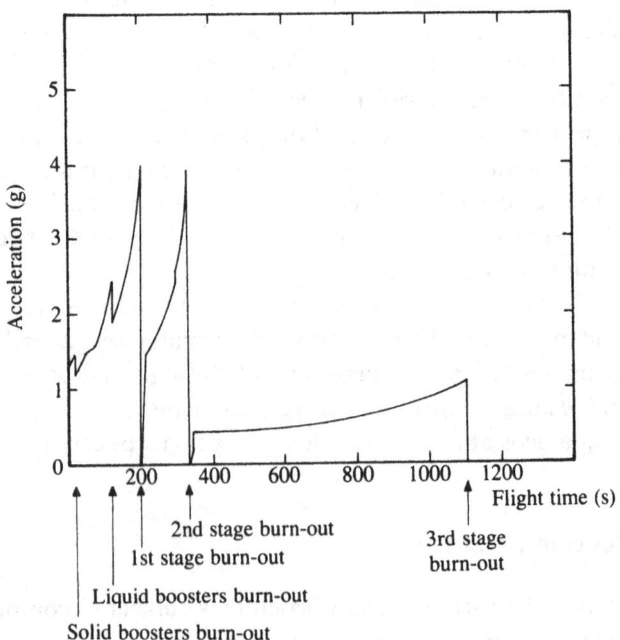

factor of the gravitational acceleration on the earth's surface. Although the thrust force of each rocket stage is relatively constant, the acceleration is not, because the stage becomes lighter as propellant is consumed. Figures 4.1 and 4.2 show acceleration profiles for two of the seven launch vehicles described in Chapter 1.

On some launch vehicles the satellite experiences low-frequency *sinusoidal vibration* during lift-off as well as in flight. A particularly violent form of sinusoidal vibration is called the *pogo effect* which stems from a resonance phenomenon in the rocket's fuel lines. A satellite is stressed to the limits in a pogo situation, and most launch vehicle designers have gone to some length to eliminate the phenomenon. Figure 4.3 shows maximum *dynamic acceleration* levels for three vehicles which still suffer sinusoidal vibrations.

At the moment of lift-off, sound wave reflexions from the ground penetrate the rocket fairing in the form of *acoustic noise*. Shortly afterwards, as the rocket velocity enters the transonic region, the atmosphere gives rise to acoustic noise through pressure fluctuations in the boundary layer surrounding the external skin of the fairing. The noise pressure is measured in dB and varies in amplitude with time and frequency (Fig. 4.4).

Combustion of propellants in the rocket engines is felt by the satellite as *random vibration*. Its measurement unit is called energy spectral

Fig. 4.3. Sinusoidal vibration levels experienced in flight.

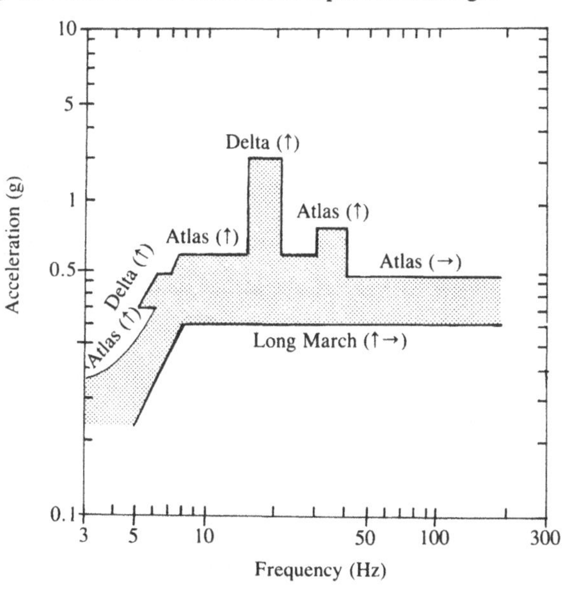

density (g^2/Hz). Random vibration covers a narrower frequency band than acoustic noise. Vibration envelopes for some launch vehicles are shown in Fig. 4.5.

Mechanical *shock* results from abrupt events in the flight sequence, such as fairing jettison, separation between rocket stages, or release of the satellite hold-down clampband at injection. The frequency spectrum of shock is obtained by Fourier analysis (Fig. 4.6).

Other Forces

In Chapters 2 and 3 we have discussed orbital perturbations due to gravitational forces acting on a satellite. In addition, solar radiation pressure gives rise to a force which affects a satellite's attitude, i.e. its orientation in space (Chapter 10). Although these forces must be regarded as important elements in the space environment, they do not

Fig. 4.4. Maximum acoustic vibration levels experienced in flight, normalized at 2×10^{-5} N/m^2.

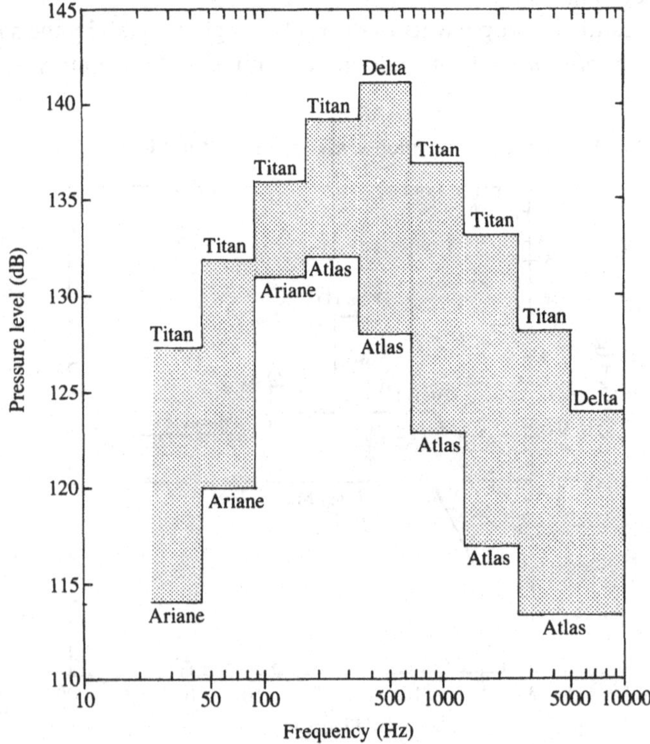

cause "bodily harm" to a spacecraft and, therefore, we will not consider them further in this context.

Atmospheric Drag

We mention drag for the sake of completeness, but as a perturbing force in a geostationary satellite's environment it is negligible. Only if the perigee of a GTO were to end up much below 120 km (e.g. due to a launcher injection error) would there be significant *orbital decay*. At that altitude the atmosphere is sufficiently dense to slow down the satellite during each passage through perigee. As a result, the apogee falls rapidly down to perigee altitude. Once the orbit is circular, it begins to contract at an accelerated pace until aerodynamic friction causes the satellite to burn up.

Radiation

X-ray and other forms of background radiation from outer space have no detrimental influence on a satellite's health. The intense light in the visible spectrum from the sun is of course mainly beneficial, since it allows solar cells to feed the satellite with electric energy. Only the ultraviolet component of sunlight is directly harmful. Under its influence, solar cells degrade in performance; thermal paints may

Fig. 4.5. Maximum random vibration levels experienced in flight.

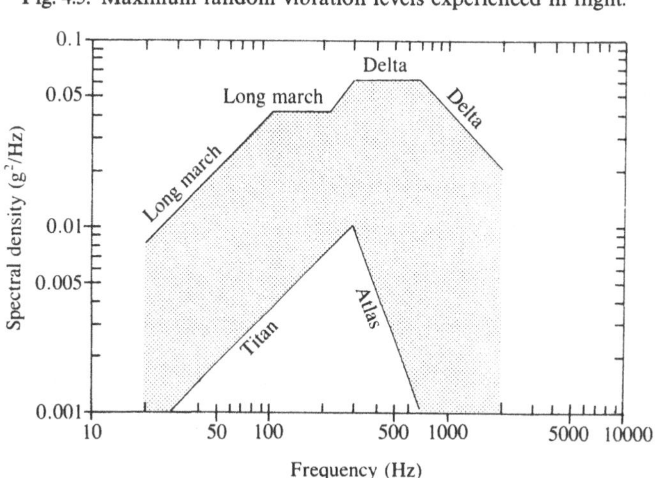

change colour so that the underlying equipment temperature exceeds its design limits; adhesives become brittle, and the conductivity of insulators increases.

After decades of spacecraft in-orbit data analysis and ground testing, scientists have successfully invented and applied new materials to protect satellites from ultraviolet radiation. Solar cell degradation remains the only serious problem, despite the advent of refined optical filtering techniques (Chapter 8).

Cosmic Particles

High-energy protons and electrons flow from outer space and the sun. *Galactic rays* consist of a sparse but energetic mixture of protons and alpha particles (helium nuclei), while the flow of protons and electrons in the *solar wind* is more dense but less energetic.

While ultraviolet radiation only affects a satellite's external surfaces, cosmic particles are sometimes sufficiently energetic to penetrate the skin and damage electronic components inside. Thus conductors, semiconductors and insulators experience excitation of their electrons

Fig. 4.6. Maximum shock spectra experienced in flight.

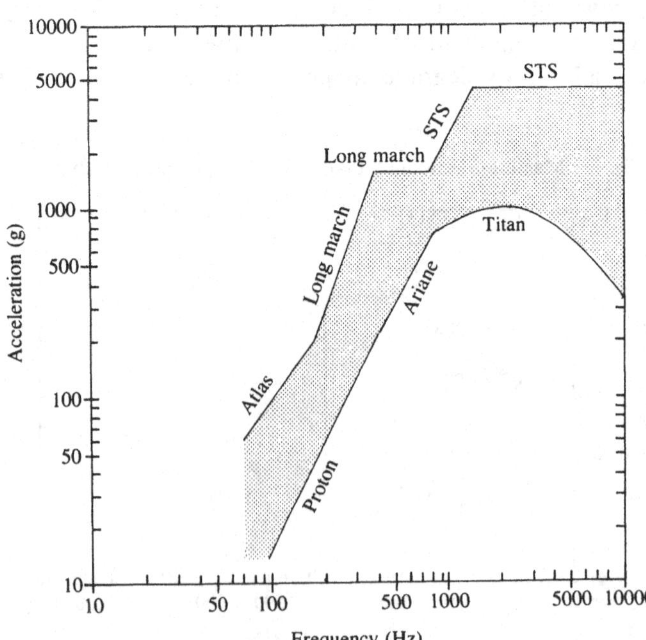

to a degree where their conductive properties change. There are temporary and permanent effects on electronic devices, especially microprocessors and memory integrated circuits. Particles also affect the optical transmission quality of glass and of certain polymers. During particularly violent solar activity, an avalanche of charged particles is emitted into space. Such events are rare, but solar cells have been known to degrade instantly by several percentage points when hit by a wave of particles.

Electrons and protons which come close to the earth are trapped along the equator by the earth's magnetic field. The high-density clouds of charged particles that result are known as *Van Allen belts.* Although the belts remain well below geostationary altitude, they have to be faced twice per orbit by a satellite travelling in GTO. There is disagreement about the extent of damage which solar cells suffer in transit, but half a percentage point per orbit has been mentioned. The prospect of severe cell degradation at the very beginning of satellite life usually prompts spacecraft controllers to fire the AKM at the earliest possible apogee in GTO.

Electrostatic Discharge (ESD)

Electrostatic discharge onboard geostationary satellites is caused by interaction with the space environment. The resulting "thunderstorms" flip electronic registers, create surges of current through electric circuits, and punch holes in insulating blankets. ESD events can set off operational alarms in spacecraft control centres; at worst, they cause essential satellite equipment to malfunction.

ESD is a common and particularly nefarious ailment which scientists and designers have tried to overcome for years. Even so, the phenomenon is only partially understood. As mentioned earlier, solar flares give rise to an avalanche of charged particles which reach the earth within hours. Many get trapped at an altitude band between 4 and 10 earth radii where they form a belt of warm (i.e. highly energetic) plasma for a short period before cooling off and dispersing. Since geostationary satellites travel at an altitude of 6.6 earth radii, they find themselves in the very centre of these disturbed plasma conditions.

In short, satellite surfaces exposed to disturbed plasma build-up electrostatic charge levels of several kilovolts relative to the environment. Moreover, different surfaces receive different charge levels, depending on the magnitude of photoemission of electrons, i.e. on whether a surface is in sunlight or in the shade. The charge level is also dependent on the kind of surface material (conductive or dielectric) used. At a given moment a breakdown of tension occurs either between the spacecraft and the plasma, or between two spacecraft surfaces. The result is a surge of current through the satellite's electric grounding plane, and sometimes an induced current in electronic circuits.

Several remedies have been proposed, such as reducing the differences in electric conductivity between exposed surfaces. These remedies have brought substantial improvements since the mid-1960s when geostationary satellites were first launched, but more research is needed into the causes of ESD as well as into alternative materials for spacecraft construction.

Cosmic Dust

Solid neutral particles from outer space are called *cosmic dust* or *micrometeorites*. There is an abundance of dust in space. Fortunately for satellites, the density decreases rapidly with increasing size. Particles with a size greater than 0.1 mm earn the title *meteorites*. At geostationary altitude they cause approximately 0.1 nm/year of surface erosion on satellites, which is usually not a problem. At very low altitudes, meteorites are attracted by the earth's gravitation and their density increases, so that the rate of erosion reaches 20 nm/year. Because geostationary satellites spend only a short time in GTO, damage due to low-altitude meteorite impact is highly unlikely, although it is a popular excuse for explaining away troublesome spacecraft anomalies.

It is interesting to note, however, that the Westar VI and Palapa B2 satellites which were recovered by the space shuttle in 1984 showed distinct traces of meteorite impact. The two satellites had been left for 9 months in orbits of a few hundred kilometres after their PKMs failed. Following their retrieval, they were kept in storage by the manufacturer, Hughes Aircraft Company. Those who had the opportunity to see the satellites could easily count some 20 visible pockmarks in the drum-shaped solar arrays of each spacecraft.

Bibliography

Maral, G. and Bousquet, M. (1986). *Satellite Communications Systems*. Chichester: John Wiley.

5

Structure

Introduction

The structure is the skeleton of a satellite. The primary design criterion for a satellite structure is that it should be rigid enough to survive the launch ascent phase while being as light and compact as possible. Low weight usually means lower launch cost, and small volume is necessary if the satellite is to fit inside the confines of the launcher fairing. Perhaps the day will come when satellites are assembled not on the ground but in orbiting space stations, in which case this fundamental design criterion will become far less onerous.

A secondary criterion is that flimsy structures such as solar panels and antennae should suffer a minimum of deformation under the influence of dynamic forces and thermal stresses in geostationary orbit. Panels have a tendency to twist during attitude manoeuvres and could actually counteract the intended movements. Parabolic antenna reflectors may change geometry due to thermal stresses as the solar incidence angle varies; the result could be a loss in antenna gain due to defocussing.

The structure must also be sufficiently stiff to prevent permanent misalignment of highly directive equipment such as antennae, thrusters and attitude sensors. In this chapter we will study the architecture of a typical geostationary satellite structure, make an inventory of materials used, follow the design logic, and discuss a mathematical modelling method.

Structure Architecture

A fairly typical geostationary satellite structure is shown in

Fig. 5.1. It is made up of a primary structure, a secondary structure and various appendages.

The primary structure is the backbone of the spacecraft. In the example shown, it consists of a central cone/cylinder assembly from which four shear walls protrude, and to which three equipment panels are attached. The cone is attached to the last stage of the launch vehicle or to the perigee kick motor, as the case may be. It relays static and dynamic loads from the rocket to the cylinder and the shear walls, and it also accommodates the bipropellant apogee motor. Inside the cylinder there is room for equipment which fits onto the concave wall, e.g. pressurant tanks. The shear walls hold most of the propulsion hardware, and the three equipment panels carry all the electrical and electronic equipment as well as the antenna farm.

The secondary structure encompasses side walls and closure panels

Fig. 5.1. Exploded view of typical spacecraft structure. Courtesy British Aerospace and INMARSAT.

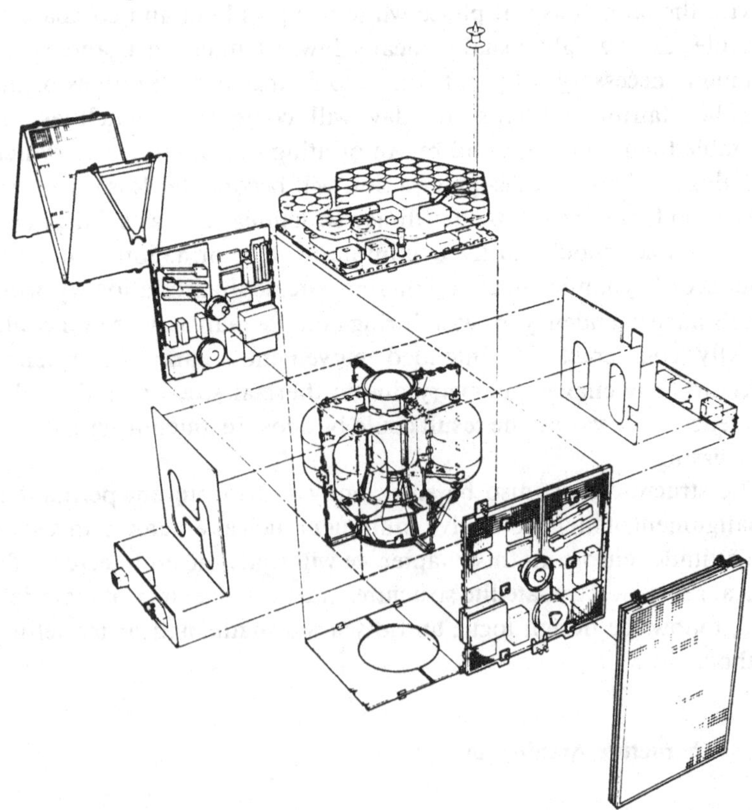

which are needed for thermal control but do not carry any significant loads.

Appendages are detachable elements fitted to the primary structure by means of bearings, hinges, struts or telescopic booms. Typical examples are deployable solar panels and boom-mounted antennae.

The satellite in Fig. 5.1 is of the three-axis stabilized variety. Structures of spin-stablized satellites have a similar architecture, except that the cylindrical solar panels form part of the secondary structure rather than the appendages.

Materials

In order to make the satellite structure as rigid and light as possible, extensive use is made of aluminium alloy. Panels and walls consist of aluminium honeycomb sandwiched between thin sheets of the same metal. Primary structure elements such as central cone/cylinder assemblies and supporting struts are frequently made of carbon fibre reinforced plastics (CFRP). Pressure vessels use titanium for extra strength, and other non-magnetic metals and plastics are employed for specialized structure applications. Materials are also chosen for their thermal conductance properties to facilitate passive temperature control of equipment (Chapter 7).

Development Philosophy

Designing a satellite structure is an iterative process. The design effort is interspersed with frequent analysis and testing activities.

To begin with, the intended satellite mission is analysed, and a conceptual structural design is developed. Chances are that a similar structure may have been used in previous satellite programmes, in which case the preliminary design work will focus on the accommodation of functional equipment onboard the existing structure.

The preliminary design lends itself to building an analytical model for stress calculations. After necessary design adjustments have been performed, the time is right to develop a preliminary finite element *mathematical model* where structural elements as well as equipment are represented by deformation *nodes* (Fig. 5.2). Computer runs of the model are initiated to simulate the displacement and rotation of each node due to static and dynamic launch loads. Evaluation of the coarse model computer runs allows stresses in critical structure elements to be analysed. Depending on the results, further re-design and re-runs may be necessary.

At this stage of structure development a real satellite structure is built, or an inherited structure is modified to resemble the intended configuration as closely as possible. Mass dummies are installed to represent satellite equipment. The structure is then subjected to *static test* and *resonance search* in order to verify mathematical predictions of static and dynamic loads. Finally, the structure is submitted to *qualification testing* (Chapter 15) to ensure that it will survive the launch and provide a satisfactory mechanical environment for the functional units mounted on it. The test results are taken into account to render the preliminary mathematical model more accurate, and the structural design is modified as required.

Fig. 5.2. Spacecraft structure mathematical model. Courtesy British Aerospace and INMARSAT.

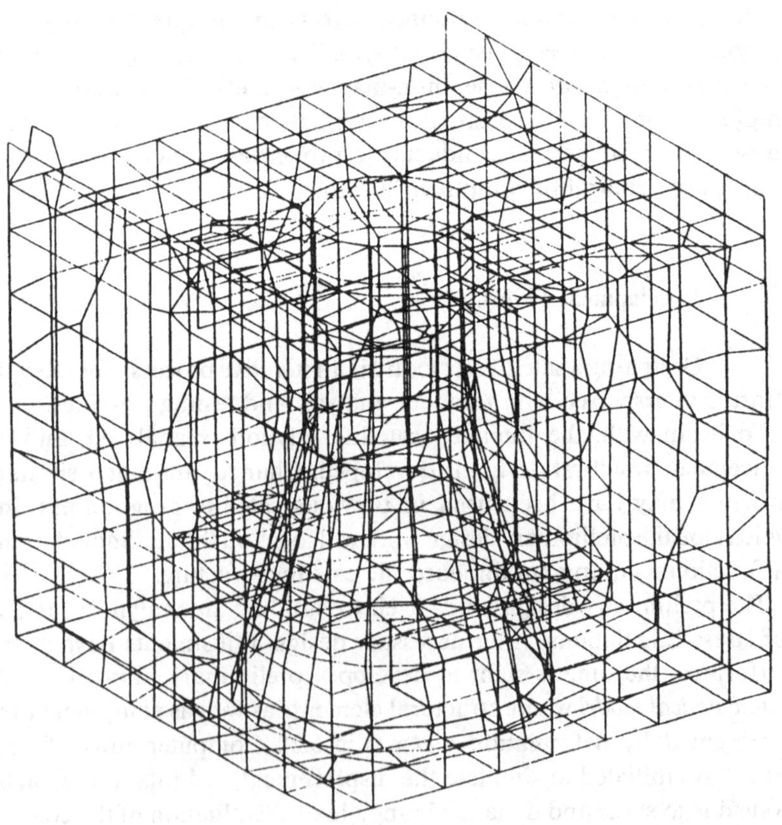

A version of the mathematical model is dispatched to the launch service agency for *coupled loads analysis* with an equivalent model of the rocket. Coupled analysis is necessary to verify that the satellite will survive not only the nominal vibration environment during launch (Chapter 4), but also loads induced by resonances with the launch vehicle, the fellow passenger and the tandem adapter, if applicable.

In summary, mathematical models assist in the iterative design process and reduce the number of test cases to be run, thereby containing the development cost within acceptable limits.

Mathematical Modelling

A satellite is composed of a variety of bars, beams, plates, curved shells and solid elements. Each is capable of supporting a combination of axial loads as well as stresses induced by shear, moment and torque. When attempting to model a complete satellite structure mathematically, it is not practicable to develop and solve a complete set of *analytical* equations, i.e. algorithms which describe deflections and stresses everywhere in all structural members. The alternative is to adopt a *numerical* analysis technique consisting of standard approximations in the form of *lumped elements* rather than continuous masses. These standard elements may be point masses, trusses, two-dimensional plates or three-dimensional shells and solids which resemble actual structural members and are joined together at discrete nodes. By assuming a certain behaviour of each standard element in response to static and dynamic loads, it is possible to compute the deflection or rotation of each node and hence the induced stresses.

Static loads may be analysed by solving simultaneous algebraic equations. Dynamic load analysis involves solving differential equations of the type:

$$m\ddot{x} + c\dot{x} + kx = F(t)$$

for each node in the mathematical model. Here, m is the mass, c is the damping constant, and k is the spring constant of the dynamic system. The time-dependent function $F(t)$ to the right of the equal-sign is the *forcing function* which emulates external forces, such as sinusoidal, random, acoustic and shock-induced forces during launch, or radiation-induced forces in orbit.

In order to model a complete structure, differential equations describing all the degrees of freedom of each finite element are grouped together into large matrices:

$$\begin{bmatrix} m_1 & & & \\ & m_2 & & \\ & & \cdot & \\ & & & m_n \end{bmatrix} \begin{Bmatrix} \ddot{x}_1 \\ \ddot{x}_2 \\ \cdot \\ \ddot{x}_n \end{Bmatrix} + \begin{bmatrix} c_1 & & & \\ & c_2 & & \\ & & \cdot & \\ & & & c_n \end{bmatrix} \begin{Bmatrix} \dot{x}_1 \\ \dot{x}_2 \\ \cdot \\ \dot{x}_n \end{Bmatrix} + \begin{bmatrix} k_1 & & & \\ & k_2 & & \\ & & \cdot & \\ & & & k_n \end{bmatrix} \begin{Bmatrix} x_1 \\ x_2 \\ \cdot \\ x_n \end{Bmatrix} = \begin{bmatrix} F_1 \\ F_2 \\ \cdot \\ F_n \end{bmatrix}$$

or simply:

$$[M]\{\ddot{x}\} + [C]\{\dot{x}\} + [K]\{x\} = [F].$$

$[M]$ is called the *mass matrix*, $[C]$ the *damping matrix* and $[K]$ the *stiffness matrix*. Matrix equations are solved by numerical finite element methods. The most widely used computer programme for solving these matrices is called NASTRAN, developed by the National Aeronautics and Space Administration (NASA).

Bibliography

Agrawal, B. N. (1986). *Design of Geosynchronous Spacecraft*. Englewood Cliffs: Prentice-Hall.
Craig, R. C. (1981). *Structural Dynamics*. New York: John Wiley.

6

Mechanisms

Introduction

Most spacecraft designers shudder at the word "mechanism". A mechanism implies moving parts which are notoriously unreliable in the space environment. Some of the causes of the high failure rate have been mentioned in Chapter 4, namely deformation due to stresses during powered flight, and evaporation of lubricant ingredients. Other common causes are changed friction coefficients in vacuum, and wear-out. These parameters are difficult to measure during testing on the ground due to access constraints in artificial vacuum chambers and the near impossibility of simulating weightlessness. As we shall see, mechanisms of one form or another are none the less necessary in the construction of modern satellites.

The Need for Mechanisms

The need is best illustrated by what they are and what they do (Figs 6.1 and 6.2):

1. *Solar array drive*, sometimes called *BAPTA*, which stands for Bearing and Power Transmission Assembly. The drive consists of an electric motor which aligns the solar paddle or wing to the sun. Used on three-axis-stabilized satellites.
2. *Solar array deployment mechanism* to release stowed solar paddles.
3. *Antenna deployment mechanism*, consisting of an antenna mounted on a boom or a yoke which is hinged at the base.
4. *Mechanically despun antenna* to keep the antenna pointed

Fig. 6.1. Examples of spacecraft mechanisms.

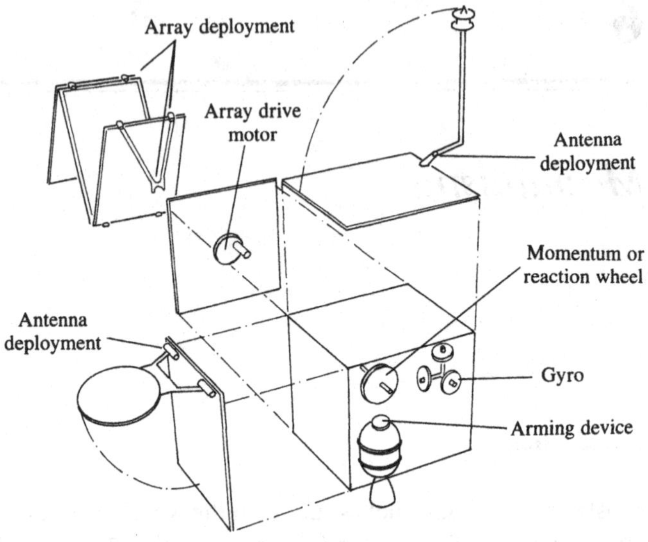

Fig. 6.2. Examples of spacecraft mechanisms.

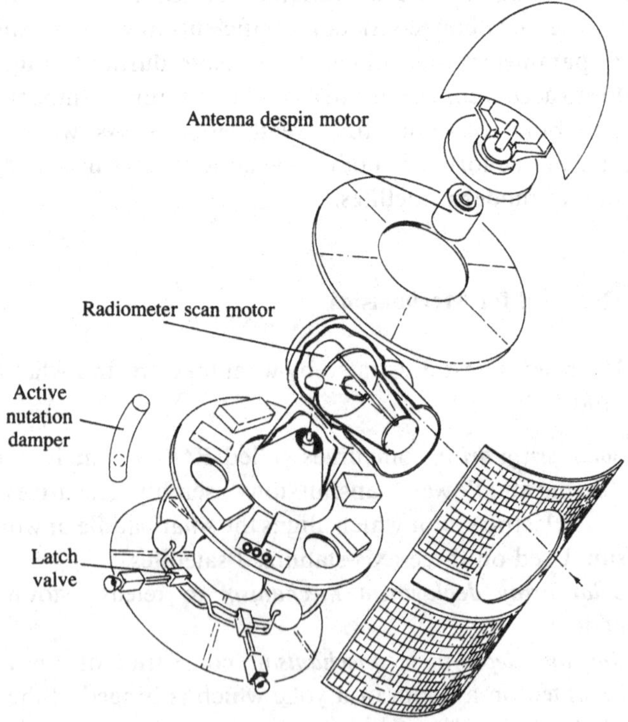

towards the earth while the rest of the spacecraft is spinning. The mechanism resembles a BAPTA for solar paddles, but the rotational speed is much higher.

5. *Momentum or reaction wheel* using gyroscopic stiffness or conservation of momentum to provide three-axis stabilization.
6. *Gyro* to provide instantaneous attitude angle and rate measurements during three-axis manoeuvres.
7. *Accelerometer*, consisting of a suspended mass whose movements give a measure of nutation.
8. *Nutation damper*, either a solid ball or a liquid moving inside a tube filled with gas to dissipate nutation energy through viscous friction.
9. *Step motor* for radiometer telescope scanning and apogee motor arming.
10. *Pyrotechnic actuators* to cut cables, open valves and release movement inhibitors.
11. *Electromagnetic valves* to open and close propellant flow in pipes.

The usefulness of each of these mechanisms will be described in later chapters.

Trade-off between Usefulness and Reliability

The use of mechanisms differs somewhat between three-axis-stabilized and spin-stabilized satellites. The former require solar paddles to be stowed during launch, deployed after injection and subsequently aligned with the sun. They also use reaction and momentum wheels as well as gyros for attitude control and measurement. The latter employ despun antennae, accelerometers and nutation dampers. Both may need to fold their antennae during launch in order to fit inside the launcher fairing.

It is often possible to increase reliability by replacing mechanisms with electronic devices. Some meteorological satellites use electronically despun phased arrays instead of mechanically despun parabolic antennae; others are equipped with electronic instead of mechanical radiometer scanners. In the past, electronic substitutes have often yielded inferior performances but, as technology advances, this drawback is likely to disappear.

At the same time, much effort is invested in making mechanisms more reliable. The use of magnetic suspension of momentum wheels instead of roller bearings is a good example. Lately, continuously operating mechanisms such as BAPTAs and wheels have in fact shown a higher reliability than single-use elements such as deployment mechanisms and pyrotechnic devices.

The trend is therefore to improve both the performance of steerable electronic devices and the reliability of mechanisms, so as to offer the spacecraft designer a wide choice of technologies for individual applications.

7

Thermal Control

Introduction

We live in a world of "room temperature technology" as far as electronics and chemicals are concerned. The ambient temperature in the technologically developed parts of the world is in the range of 0–30°C. Nearly all space-qualified materials are derived from earth-bound applications, be it electronic components, electrolytes, lubricants, paints or adhesives. Equipment on earth having a tendency to run too hot or too cold may be readily brought back to acceptable operating temperatures through heat exchange with the atmosphere.

In space, however, there is only extreme cold and extreme heat. If no action were taken, passive satellite equipment would adopt temperatures from typically −200 to +150°C, while active electronics might reach temperatures of several hundred degrees. To make things worse, a satellite will occasionally dive into the earth's shadow and emerge again into sunlight, such that fierce thermal stresses develop within the spacecraft. Because vacuum prevails, there is no heat convection. The only heat exchange is through radiation and conduction, and they are poor substitutes for convection when it comes to creating some kind of temperature balance.

It is therefore necessary to create approximate room temperature inside a spacecraft where most of the electronics and chemicals are found. This is the task of spacecraft thermal control. *Passive control* using thermal blankets, paints and other surface treatments go a long way to create the right operational environment, but sometimes it is necessary to resort to *active control* with the aid of electric heaters, louvres and heat pipes.

Basic Theory

Take an isothermal body such as a solid, homogeneous sphere and launch it into space. Depending on its colour, it will absorb a certain fraction of sunlight energy and radiate it back into space. The absorbed light arrives mostly in the *visible* spectrum (0.4–0.7 μm), while the radiated energy is in the *infrared* (IR) spectrum (10–14 μm). The size of the illuminated surface determines the amount of *absorbed* energy, whereas thermal energy is *emitted* in all directions.

In a state of thermal equilibrium, we have:

Absorbed power = emitted power

or

$$\alpha A_a C = \varepsilon \sigma A_e T^4, \tag{7.1}$$

where α = the absorbtivity coefficient ($0 < \alpha < 1$), ε = the emissivity coefficient ($0 < \varepsilon < 1$), A_a = the absorbing surface area (= πr^2 for a sphere), A_e = the emitting surface area (= $4\pi r^2$ for a sphere), C = the solar constant (average 1353 W/m²), T = body temperature in degrees Kelvin and σ = Stefan-Boltzmann's constant (5.67×10^{-8} W/m² K⁴).

As illustrated above for a sphere, $A_a \neq A_e$ in the general case.

From Eqn 7.1 we obtain for a sphere:

$$T = \left(\frac{\alpha}{\varepsilon} \cdot \frac{C}{4\sigma}\right)^{1/4}. \tag{7.2}$$

Fig. 7.1. Equilibrium temperature of an isothermal sphere in space as a function of its absorbivity α and emissivity ε.

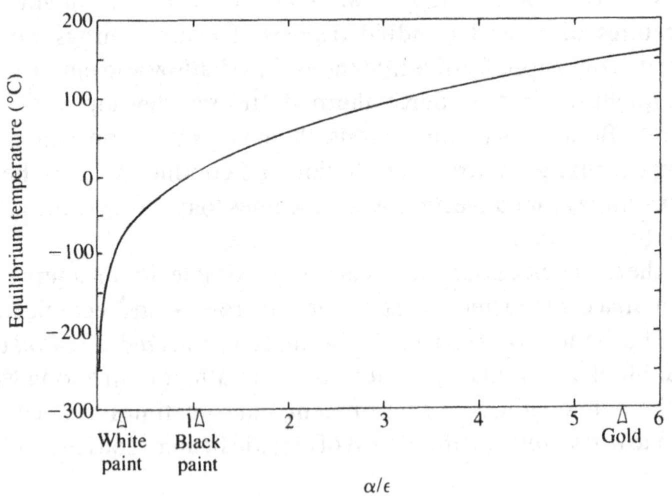

Table 7.1.

	α	ε	α/ε
White paint	0.20	0.90	0.22
Black paint	0.95	0.90	1.05
Aluminium paint	0.25	0.25	1.00
Gold	0.25	0.045	5.50
Graphite epoxy	0.95	0.75	1.25
Fibreglass	0.90	0.90	1.00
Aluminized Kapton	0.50	0.60	0.83
OSR (optical solar reflectors)	0.08	0.80	0.10
SSM (second surface mirrors)	0.15	0.80	0.19
Solar cells	0.80	0.90	0.90

(see Fig. 7.1). Some examples of absorbtivity and emissivity coefficients can be found in Table 7.1. Note that the temperature of an isothermal body in space is entirely predictable, as long as its physical shape and surface treatment are known. A thermal control designer should in principle be able to achieve a specified temperature distribution within a satellite by simply applying a variety of paints, mirrors or blankets to different parts of the spacecraft.

In reality, the job of a thermal designer is complicated by the fact that a satellite is not a solid, isotropic body but a hollow structure made up of materials with different radiative and conductive properties. What is more, some of the units dissipate heat continuously or intermittently, and the solar incidence angle changes by the hour and by the season. The complexity of thermal design and prediction turns out to be so great that the designer must resort to numerical analysis methods similar to those described in Chapter 5 for structures.

Passive Thermal Control Materials

Multilayer superinsulation blankets consist of alternating layers of aluminized Kapton and Dacron mesh. The purpose of the mesh is to separate adjacent layers of Kapton sheet so that no temperature leakage occurs through conduction. The outermost layer is often covered with a film of indium-tin oxide (ITO) which makes the blankets electrically conductive, thereby preventing electrostatic charging and associated discharges. The blankets have low α and ε, i.e. they are relatively inert to variations in heat flux. On three-axis stabilized satellites they cover most of the four external walls which receive direct sun illumination in

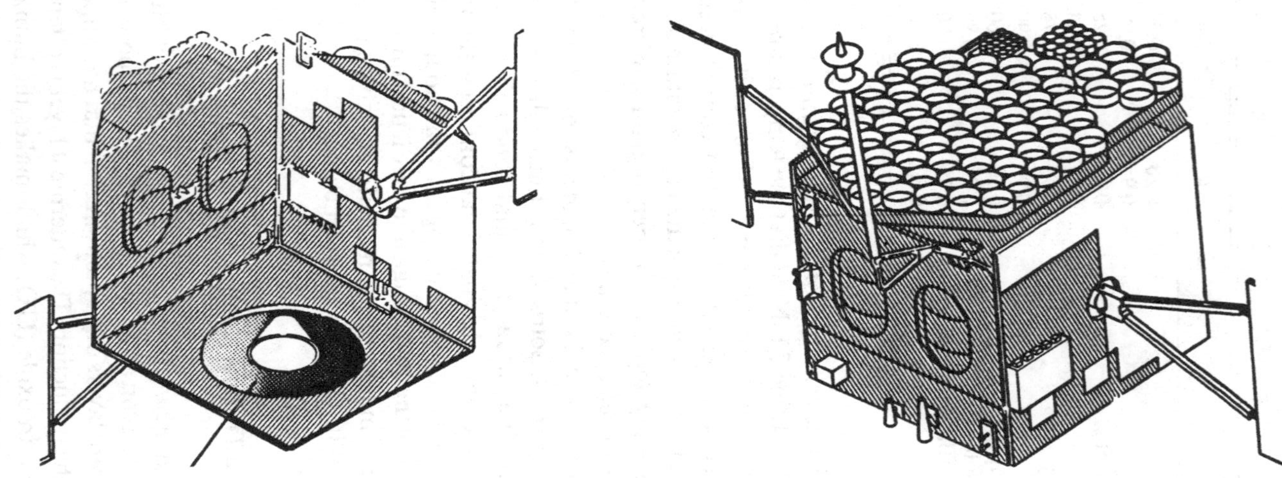

Fig. 7.2. Multi-layer insulation blankets (shaded areas). Courtesy British Aerospace and INMARSAT.

the course of a day (Fig. 7.2), thus protecting electronic and propulsion equipment inside from extreme temperature variations.

Blankets are also wrapped around propulsion subsystem elements to avoid propellant freezing. Walls surrounding the apogee kick motor are covered with blankets to shield the satellite from the immense heat emitted by the burning motor.

Optical solar reflectors (OSR) are made of fused silica glass with silver backing. With their low α and high ε, the reflectors are ideally suited for dissipating heat from units which work best at low temperatures, such as power amplifiers and batteries. On three-axis as well as spin-stabilized satellites, OSRs are found on the outside of north and south walls which, like the North and South Poles, receive little solar illumination during the year.

Second surface mirrors (SSM) resemble OSR in their use and appearance, except that they are made of flexible plastic sheets instead of glass. They are easier to handle than OSR but degrade more rapidly with age (increasing α) due to charged particle bombardment and ultraviolet radiation.

Aluminium doublers (Fig. 7.3) provide a large equipment mounting footprint for improved thermal dissipation through conduction. High-power amplifiers are frequently mounted on doublers rather than directly onto north and south walls.

White paint is used primarily on parabolic antenna surfaces to reduce temperature fluctuations and thereby prevent mechanical distortion.

Black paint covers many electronics boxes inside a satellite. The high values of both α and ε help maximize the heat exchange with other onboard equipment.

Fig. 7.3. Aluminium doubler between a travelling wave tube and a spacecraft external wall.

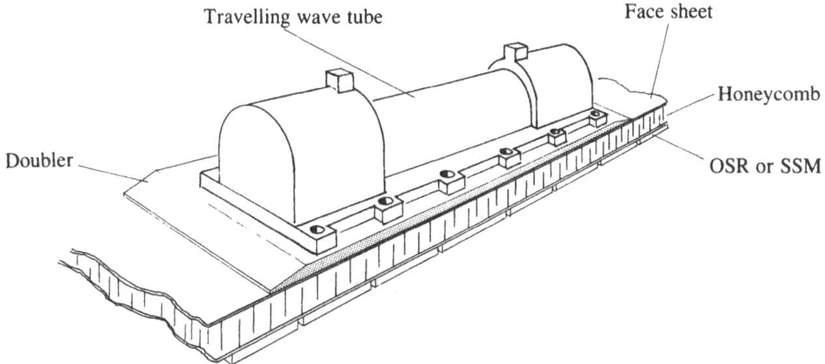

Solar arrays, while not primarily a thermal control medium, contribute to the temperature environment inside spin-stabilized satellites. If array switching regulators are used to control the main bus voltage (Chapter 8), any excess power has to be dissipated by the arrays themselves which therefore tend to run relatively hot. Conversely, if shunt regulators are employed to absorb excess power, the arrays will operate at colder temperatures. The choice of main bus regulation technique will therefore influence the overall heat balance inside the satellite.

Active Thermal Control Equipment

Electrical heaters are resistors which heat up when fed with an electric current. They are frequently attached on tanks and pipes to prevent propellant freezing. Heaters are sometimes mounted on travelling wave tube amplifiers (Chapter 12) as *compensation heaters* or *simulation heaters* to create thermal balance between adjacent operating and idle amplifiers. They are also used on the inside of radiometer IR sensor cooler walls (Chapter 13) to evaporate condensed outgassing products.

Heat pipes are a form of primitive refrigerator. The ability of a liquid to evaporate and condense while absorbing and emitting latent heat is exploited to transport heat from high-power amplifiers to nearby radiators. An amplifier is mounted onto one closed end of a heat pipe, while the other closed end radiates into space (Fig. 7.4). The heat dissipated by the amplifier evaporates the liquid inside the pipe. The vapour moves under its own pressure to the cold end of the pipe where condensation takes place and heat is released. Capillary action makes

Fig. 7.4. Heat pipes embedded in spacecraft wall. Pipe ends have been cut open for clarity.

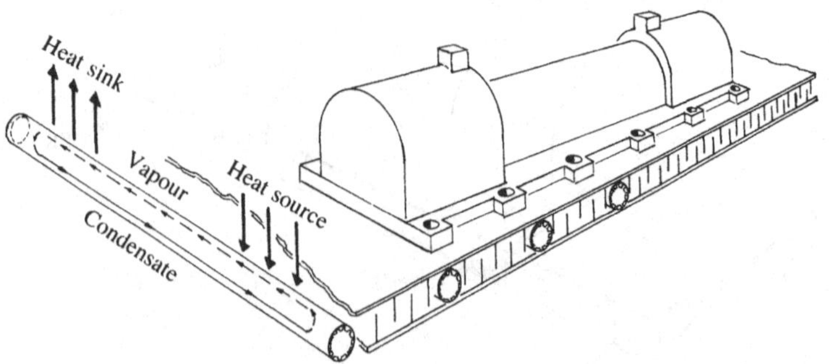

the condensed liquid travel back in grooves cut along the pipe wall to the amplifier. The evaporation/condensation process is then repeated.

Heat pipes are filled with ammonia, freon, methanol, ethanol or water, depending on the temperatures to be transported. The pipes come in lengths of up to 1 m. Their ability to divert heat is superior to any passive thermal control method, but a price must be paid in terms of increased mass.

Louvres can be made to open or close by feedback control using a bimetal temperature sensor (Fig. 7.5). They are sometimes used to provide a finer temperature regulation of batteries than simple optical solar reflectors or second-surface mirrors.

Mathematical Modelling

When analysing the performance of a thermal control subsystem, thermal designers use a *finite differences method*, which is a variant of the finite element method used in the design of structures (Chapter 6).

The spacecraft is again represented by a set of nodes (Fig. 7.6). A node will exchange heat with another through conduction and radiation. The rate of conduction depends on the length of the conductive path and the choice of material, while the radiation flow is a function of the geometrical view factor. The energy input to a given node "*i*" comes from direct solar flux θ_s, reflected solar flux θ_a from the earth (*albedo*) and infrared radiation flux θ_e from the earth (*earthshine*), as well as from

Fig. 7.5. Louvres mounted on spacecraft external wall. Emissivity $\varepsilon = f(\varepsilon_1, \varepsilon_2, \theta)$.

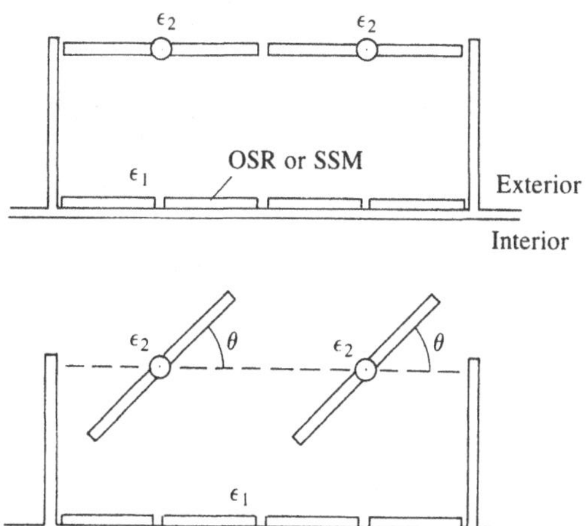

Fig. 7.6. Spacecraft thermal mathematical model. Courtesy British Aerospace and INMARSAT.

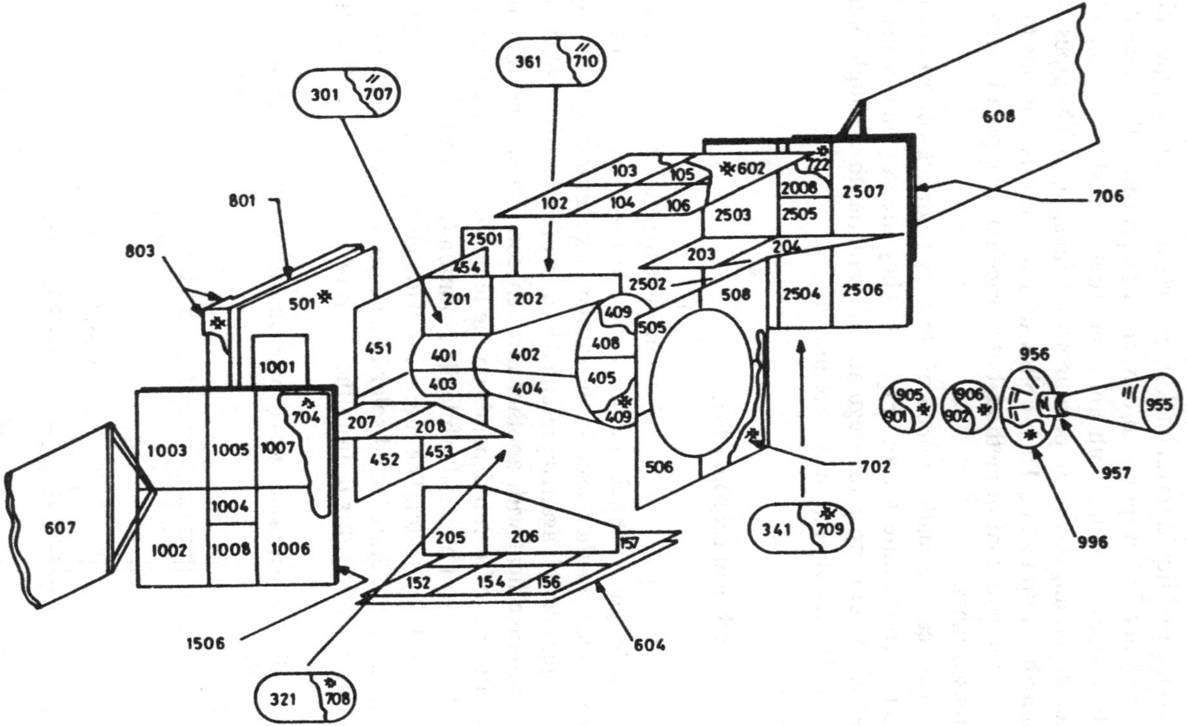

equipment energy dissipation P within the node. The output is the flux lost through radiation R and conduction C as well as the energy stored at the node during a time dt. Thus, for each node:

Absorbed flux = radiated flux

or

$$\alpha_i A_i \phi_i \theta_{si} + \alpha_i A_i \theta_{ai} + \varepsilon_i A_i \theta_{ei} + P_i$$
$$= \sum_{j=1}^{n+1} \sigma R_{ij}(T_i^4 - T_j^4) + \sum_{j=1}^{n} C_{ij}(T_i - T_j) + m_i c_i \frac{\mathrm{d}T_i}{\mathrm{d}t}$$

where A_i = the radiant area of node i (m^2), ϕ_i = the solar aspect coefficient (projection of illuminated area/A_i), m_i = equipment mass of node i (kg), c_i = specific heat of node i (J/K kg), $m_i c_i$ = thermal capacity (J/K), T_i, T_j = temperatures at nodes i and j (K), C_{ij} = conductive coupling between nodes i and j (W/K), R_{ij} = radiative coupling between nodes i and j (m^2), P_i = internal power dissipation at node i (W), θ_i = energy flux onto node i (W/m^2) and σ = Boltzmann's constant (see Eqn 7.1).

These equations are solved for various transient and equilibrium cases such as (1) sun angles during transfer orbit, (2) apogee motor burn, (3) different modes of spacecraft stabilization, (4) different spacecraft geometries due to deployment of mechanisms, (5) sun angle at equinoxes and solstices, (6) eclipse transits, (7) albedo and earthshine variations, and (8) intermittent operation of heat-dissipating equipment. The initial model inputs and outputs need not be entirely representative of the true space environment, but rather of the environmental approximation found in a *solar simulation chamber*. After calibrating the model against solar simulation test data of the real satellite, the designer may subsequently enter boundary conditions which simulate the authentic space environment, and can expect to obtain reasonably authentic results.

Thermal testing of satellites is described in Chapter 15.

Bibliography

Agrawal, B. N. (1986). *Design of Geosynchronous Spacecraft*. Englewood Cliffs: Prentice-Hall.

Redor, J. F. (1973). *Introduction to Spacecraft Thermal Control*, ESTEC Internal Working Paper No. 768. Noordwijk: European Space Technology Centre.

8

Power Supply and Conditioning

Introduction

By the time the light emanating from the sun reaches the geosynchronous earth orbit, its energy content has been reduced to approximately 1350 W/m^2. Hence, if all of the light that falls on a 1 m^2 surface could be converted to electric power, it would be just enough to run an electric iron. In practice, only 10–14% can be thus converted with today's technology. The engineering challenge facing a satellite designer is to draw on this meagre solar energy to operate and heat an entire spacecraft, and to transmit radio waves with sufficient signal strength to be received intelligibly on the earth some 36 000 km away.

This chapter describes how electric energy is generated, conditioned and stored onboard geostationary satellites.

Subsystem Architecture

The main building blocks are the solar array, the battery and the loads (Fig. 8.1). The solar array is made up of strings of photovoltaic cells cemented onto solar panels. The battery is charged by the array during sunlight so as to provide power in eclipse when the array is idle. The loads are the users of electric power, such as transmitters, receivers, microprocessors, electric motors and heaters. The battery can also be considered a load while it is being charged.

A satellite power subsystem contains switches to turn on or off various load combinations and to protect the battery from overcharge or depletion. Some of the switches are triggered automatically when certain conditions arise, while others are activated by telecommand. A manual override of automatic switching functions by telecommand is usually provided to cover unforeseen contingencies.

Power Generation

The monocrystalline silicon solar cell was invented in 1954, only 3 years before the launch of the first artificial satellite. Without it, the exploitation of space might have been delayed by decades. A cell is made up of a thin *n*-layer diffused into a *p*-type substrate to form an *n–p* wafer. The *n*-layer faces the incoming light which displaces electrons, leaving behind electron "holes". As other electrons move to fill the holes, an electric current is generated.

A cross-section of three interconnected cells is shown in Fig. 8.2. Energy conversion takes place in the silicon wafer. Electric current is collected through a lead from the *n*-layer of one cell to the *p*-layer of the adjacent cell. The wafers are doped with gallium or boron for longer life. Each cell is protected from excessive electron and proton radiation damage by a cover glass made of fused silica. By doping this silica with cerium, the risk of electrostatic build-up and subsequent discharge is reduced (Chapter 4). The external glass surface is covered with an anti-reflexion film of tantalum pentoxide (Ta_2O_5), while the internal surface is coated with an ultraviolet filter to prevent the underlying transparent adhesive from becoming opaque.

Fig. 8.1. Basic elements of a spacecraft power supply and conditioning subsystem.

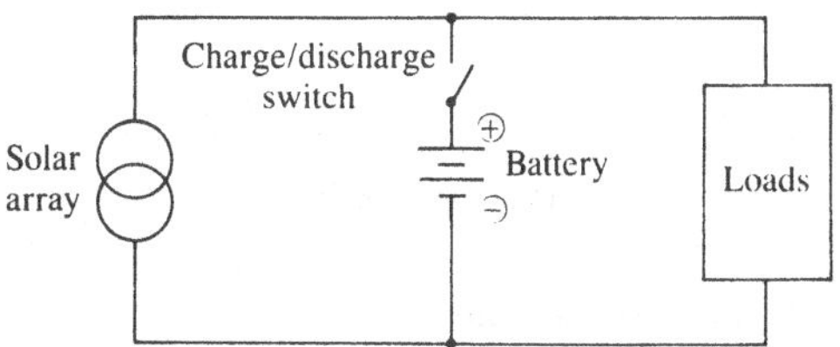

Fig. 8.2. Solar cell structure and interconnection.

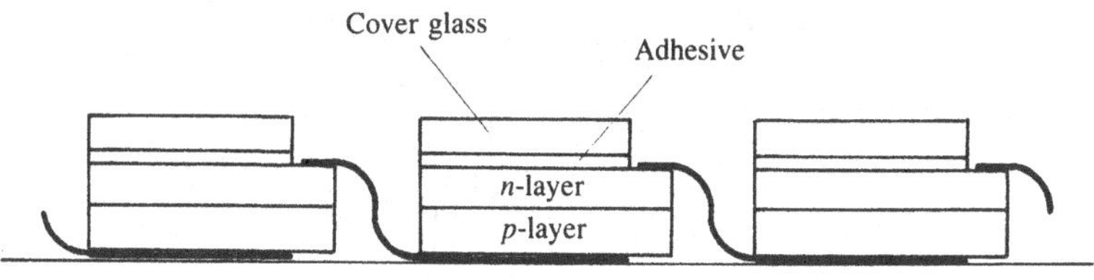

Solar cells come in sizes from 2 × 2 cm to 4 × 8 cm. Their thickness has traditionally been 200 μm, but the trend is towards ultra-thin cells measuring 50 μm or less. Thinner cells are not only lighter but also more efficient (typically 14% as opposed to 10%).

A solar cell is basically a *current generator*, i.e. the output of current increases with illumination intensity while the voltage across the cell stays relatively constant. The current output is also a function of cell temperature (Fig. 8.3). By stringing cells together in parallel, the current output is increased. Voltage is built up by connecting cell strings in series. In order to meet specified power supply requirements, a solar array is laid out as a matrix of cells connected in series as well as in parallel (Fig. 8.4). The cell layout on actual solar panels is often made more complex due to mission-related requirements. For example, steps may be taken to minimize power loss in case of accidental open or closed circuits within individual cell strings, or it may be decided to charge the battery from a separate mini-array embedded in the main

Fig. 8.3. Solar cell output characteristics.

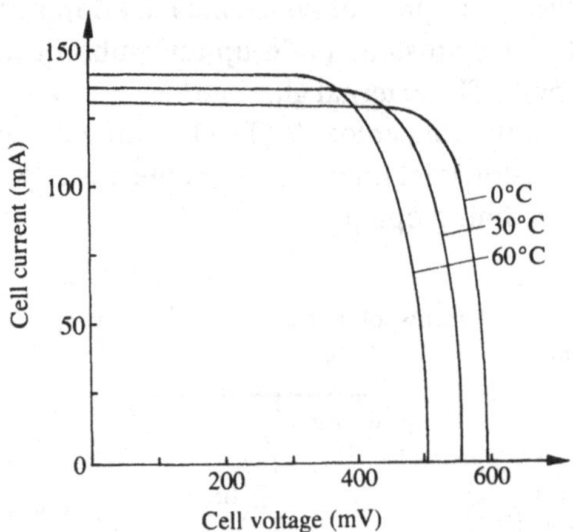

Fig. 8.4. Solar array wiring diagram.

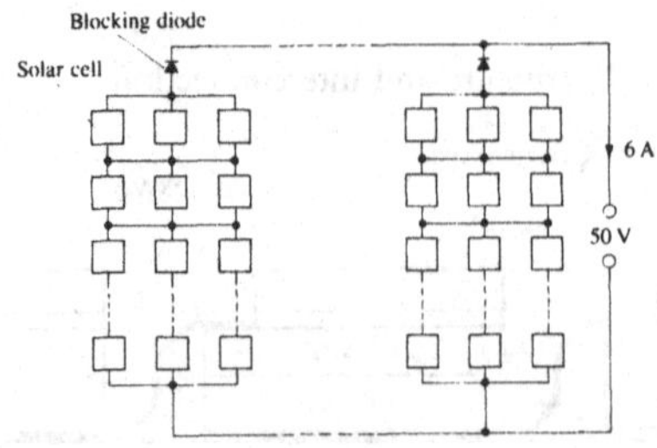

array. Some cylindrical arrays mounted on spin-stabilized satellites have cut-outs for optical sensors, forcing a re-layout of the other arrays to provide constant electrical output as the satellite rotates.

The diodes shown at the top ends of the strings in Fig. 8.4 are called *blocking diodes*. They ensure that reverse current from sunlit cells do not damage idle cells obscured by the shadow of the spacecraft main body.

The geometry of solar panels differs depending on the spacecraft stabilization method. On three-axis-stabilized (or "body-stabilized") satellites the panels form deployable *solar wings* or *paddles* (Fig. 8.5). The wings are aligned with the earth's north–south axis. Each wing is made to face the sun by an electric step motor which turns at a rate of one revolution per 24 h.

A wing is normally made of rigid panels which are hinged together. They are folded up against the sides of the spacecraft during launch such that the arrays on the outboard panels face outward. This way the satellite receives enough electric power while in sunlight to survive from the moment the launcher fairing is jettisoned until the wings are deployable in GEO several hours or days later.

Rigid, hinged panels are likely to give way to flexible array substrates in the future. These are either compressed like an accordion or wound

Fig. 8.5. Deployment of solar wings of a three-axis-stabilized satellite. Courtesy of British Aerospace and INMARSAT.

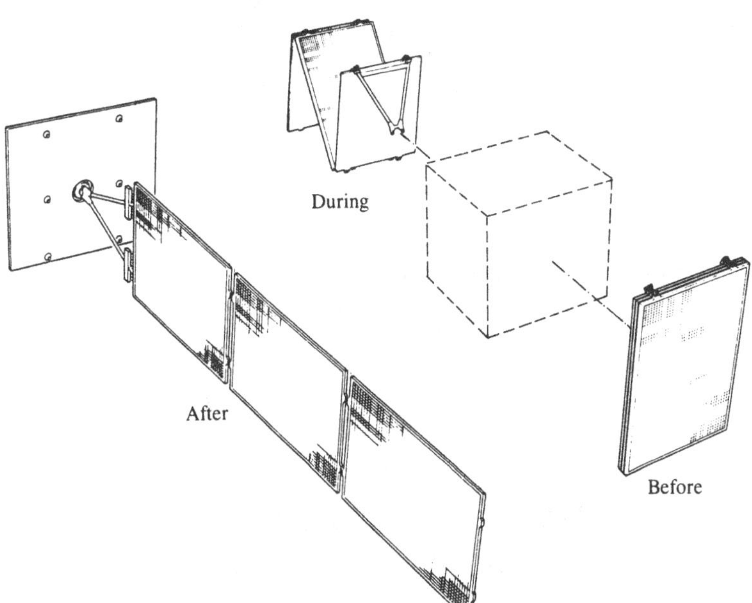

up around cylinders before launch. Once the satellite has been injected into the geostationary orbit, the arrays are unfurled to provide adequate power without upsetting orbit and attitude manoeuvres.

Spin-stabilized satellites use cylindrical panel substrates (Fig. 8.6) which also form the main secondary structure (see Chapter 5). Because half of the array is in shadow at any one time, and because the sunlit half is mostly illuminated on a slant (Fig. 8.7), cylindrical arrays are less efficient than wing-mounted arrays. Take a cylinder of height h and diameter d exposed to perpendicular illumination. The overall array area equals $\pi d \cdot h$, whereas the effective illuminated area is only $d \cdot h$.

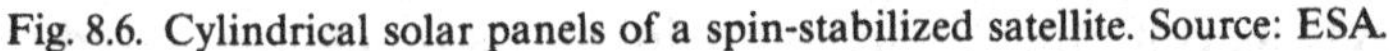

Fig. 8.6. Cylindrical solar panels of a spin-stabilized satellite. Source: ESA.

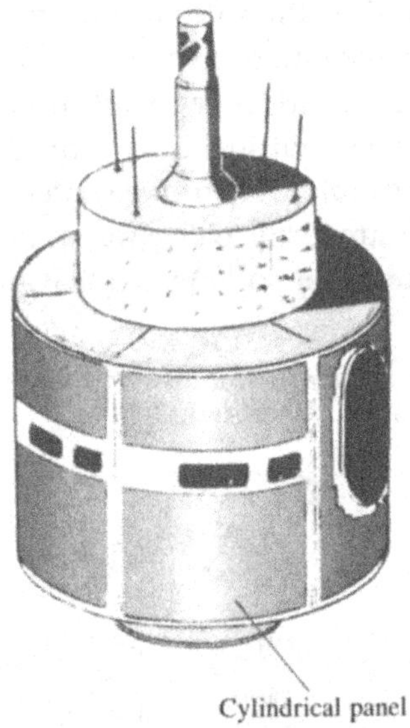

Fig. 8.7. Effective illuminated area of a cylindrical solar array.

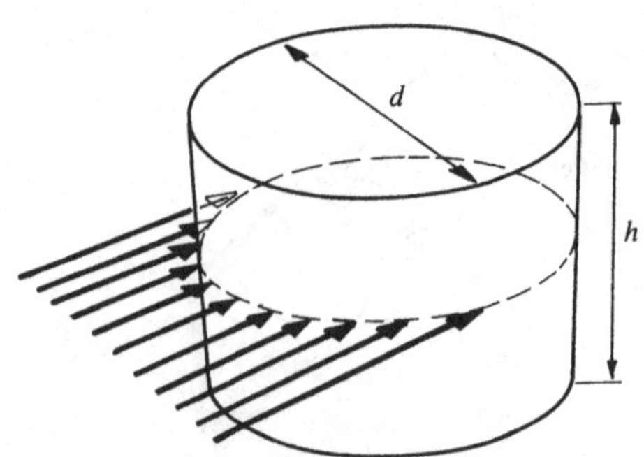

Consequently, a spin-stabilized satellite array surface has to be π times as large as the equivalent three-axis wing for the same power output performance. This affects both the mass and the cost of a satellite.

Hughes Aircraft Company in the US, who builds spin-stabilized satellites almost exclusively, has managed to increase the power output with minimum mass penalty by adding a cylindrical array-covered skirt (Fig. 8.8). During launch the skirt covers the body array like an outer skin. Following injection, the skirt is telescoped downward to expose the body array. Together they provide twice the power output of a traditional spin-stabilized satellite of similar dimensions.

Batteries

The batteries installed onboard most geostationary satellites employ one of two electrode technologies: *nickel-cadmium* (NiCd) or *nickel-hydrogen* (NiH_2). The former have well-established performance characteristics, whereas the latter are of more recent vintage and promise longer life for lower mass. A NiCd battery is illustrated in

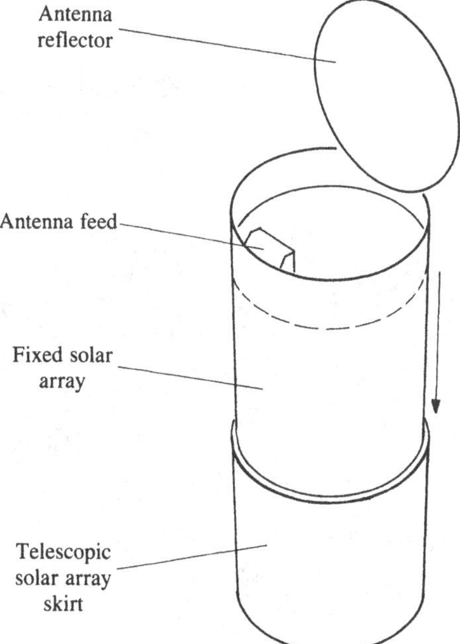

Fig. 8.8. Telescopic extension of cylindrical solar array.

Fig. 8.9. NiCd cells have a box-shaped geometry. They are stacked wall-to-wall in sufficient number to yield the specified ampere-hour capacity, and are clamped together between end walls. NiH_2 cells are subject to internal gas pressure and are therefore cylindrical in shape with spherical end caps. The cells are clustered to give the desired battery capacity.

The chemical reactions taking place inside a NiCd battery cell during charge and discharge are:

$$Cd(OH)_2 + 2Ni(OH)_2 \underset{\text{discharge}}{\overset{\text{charge}}{\rightleftharpoons}} Cd + 2NiOOH + 2H_2O.$$

Similarly, for a NiH_2 cell:

$$Ni(OH)_2 \underset{\text{discharge}}{\overset{\text{charge}}{\rightleftharpoons}} NiOOH + \tfrac{1}{2}H_2.$$

Batteries are typically made up of some 25 cells, each with an average discharge voltage of 1.2 V. The storage capacity is in the range of 30–50 Ah, though the allowable *depth-of-discharge* (DOD) is limited to 60–80%. A battery weighs 30–40 kg depending on size and technology.

Correct battery discharge management is of great concern during orbital operations. If the permitted DOD is exceeded, the battery

Fig. 8.9. Nickel-cadmium battery. Courtesy of British Aerospace and INMARSAT.

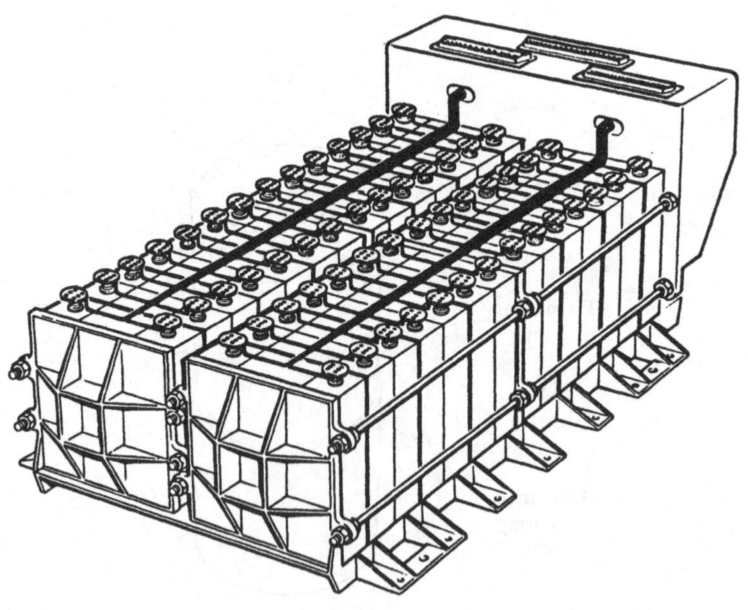

lifetime may be shortened. The DOD is measured by monitoring the battery output voltage through telemetry. This measurement is inaccurate, however, due to the relatively flat voltage profile during normal discharge. Spacecraft controllers therefore supplement the voltage monitoring with other information such as battery temperature and, in the case of NiH_2, pressure. Dead reckoning, i.e. keeping a log of onboard power consumption during eclipse, is also used. Similar management techniques are employed to charge batteries without overcharging them during periods of sunlight.

Two or three batteries are usually installed to provide adequate capacity as well as redundancy. They are *mission critical* for those geostationary satellites which must offer a full service day and night, including eclipse periods. Communications and meteorological satellites fall into this category. For television broadcast satellites, however, batteries are not necessarily mission critical if the eclipse transit occurs at a time when no television transmissions are foreseen. For example, a European satellite located at 30°W longitude goes into eclipse at 01.25 GMT at the very earliest, by which time all TV transmissions within Europe would have ceased. Such a satellite can survive on a single battery for housekeeping purposes and save 40–80 kg of mass, compared with a spacecraft requiring redundant batteries to serve a user community 24 h a day.

Power Conditioning

The main power line in a spacecraft is called the *bus* (not to be confused with the word "bus" used as a synonym for "platform"). The bus is like a water main which originates at a source (the solar array) and branches out to many households (loads). A water tower (the battery) is connected in parallel with the source such that it may be filled (charged), as well as drained (discharged) at times when the source dries up (eclipse).

When a household starts drawing water, the pressure (voltage) drops slightly. If all the households were to turn on all their taps at the same time, the pressure from the source would be inadequate to meet the demand on water flow (current), and the water tower would have to step in to restore the pressure and hence the flow of water. There are other ways to regulate the pressure, however, such as controlling the overflow gates at the source reservoir and dissipating excess water into drainage channels. Pressure may also be increased, decreased or even reversed with the aid of pumps. In electrical terms, these actions are called *power conditioning*.

A *regulator* is an active electronic device which ensures a constant voltage or power output when the input varies within reasonable limits. A regulator also provides a constant output when the loads vary downstream. A *converter* is the direct current (DC) equivalent of an alternating current (AC) transformer. Older converters actually chopped the input DC voltage into an AC voltage, passed the AC through a transformer, and finally rectified the output back to DC. Contemporary DC/DC converters use pulse width modulation (PWM) in the chopping process whereby the ratio between the pulse widths of "ones" and "zeros" determines the output voltage after rectification. In practice, regulation and conversion usually go hand in hand.

In an *unregulated bus*, the array output ports are connected directly across the battery terminals, and the bus adopts whatever voltage the battery happens to maintain (Fig. 8.10). The bus voltage therefore varies widely, especially around eclipse, and each load must be equipped with its own regulator and converter. The unregulated bus concept is simple and reliable, but all the decentralized regulators add up in terms of mass.

Fig. 8.10. Unregulated power bus concept.

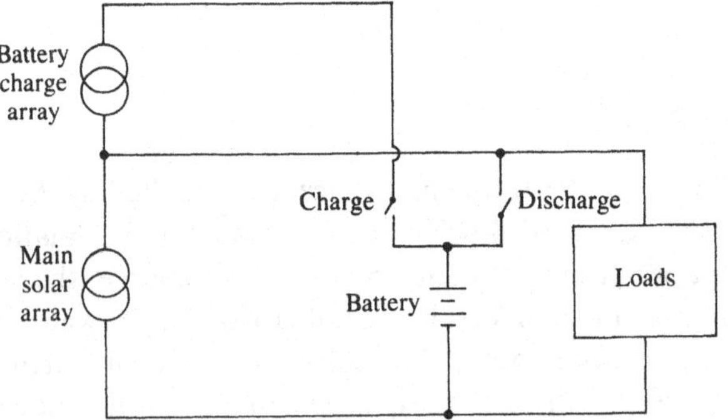

Fig. 8.11. Fully regulated power bus concept.

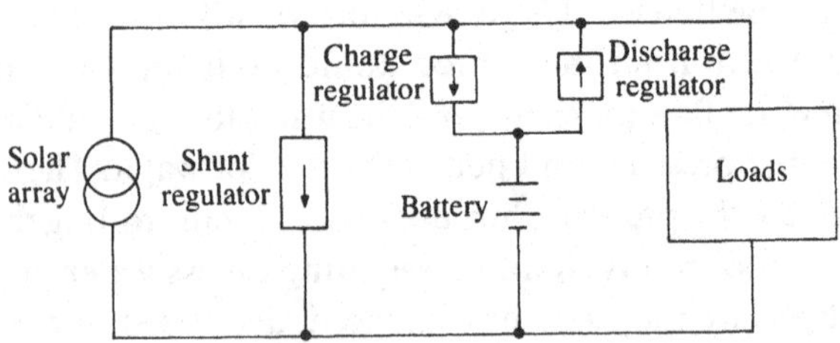

A *regulated bus* concept is shown in Fig. 8.11. here a central regulator is placed upstream near the source, thus eliminating the need for individual load regulators. The central regulator may exist either in the form of a parallel shunt or a serial array switching device. Variations on these design principles exist. A *shunt regulator* takes excess power from the solar array and dissipates it, whereas an *array switching regulator* measures the instantaneous load demand and switches array sections (groups of strings) in and out to balance the demand. The batteries are equipped with discharge regulators in order to maintain bus regulation during eclipse.

A hybrid between the two regulation methods is the *sunlight regulated bus* (Fig. 8.12), which is regulated in sunlight but not in eclipse. It dispenses with battery discharge regulators while protecting the loads from high voltage extremes occurring just after eclipse when the solar arrays are very cold. In other words, the sunlight regulated bus reduces the voltage excursions compared to an unregulated bus and simplifies the design of individual load regulators.

Power Balance

The power output P from a solar array varies with the cosine of the solar aspect angle θ_s':

$$P = P_0 \cos \theta_s' \tag{8.1}$$

with P_0 being the maximum achievable output (Fig. 8.13). The cut-off at incidence angles close to 90° is caused by total reflexion.

P is also a function of the incident light intensity I. As the earth travels along its slightly elliptic orbit, I varies according to:

$$I = 1354 + 45 \cos (DD\ 360/365.25) \qquad (\mathrm{W/m^2})$$

where DD is equal to the number of days from *perihelion* (the "perigee" of the earth's orbit around the sun). The earth travels through the

Fig. 8.12. Sunlight regulated bus concept.

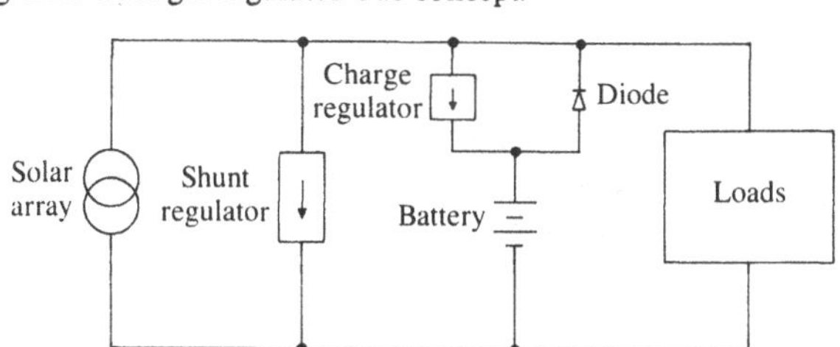

perihelion on 3 January, but 1 January may be used for convenience without significant loss of accuracy. We may normalize the calendar-dependent intensity with respect to the intensity at the equinoxes (1353 W/m^2) as follows:

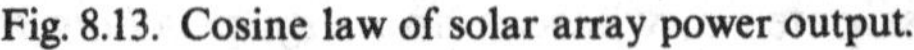

Fig. 8.13. Cosine law of solar array power output.

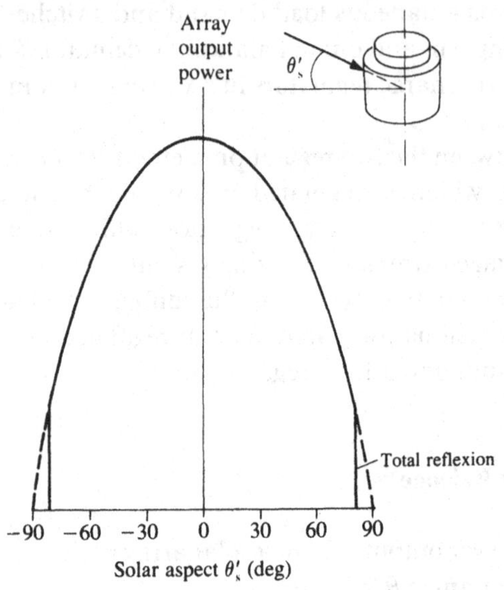

Fig. 8.14. Annual variations in solar incidence angle and intensity.

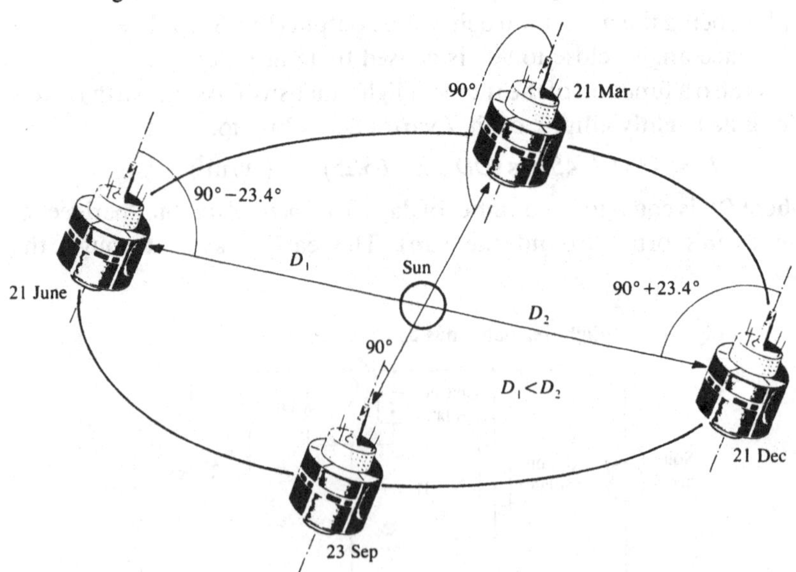

$$J = I/1353. \tag{8.2}$$

The incidence angle θ_s' undergoes two cyclic changes per year between +23.44° and −23.44° due to the tilt of the earth's spin axis with respect to the ecliptic plane normal (Fig. 8.14). Thus:

$$\sin\theta_s' = \sin 23.44° \sin\{(DD + 101)\,360/365.25\}. \tag{8.3}$$

Moreover, the array degrades by some 30% over 10 years due to radiation damage. If D is the degradation factor, we have:

$$D \simeq 0.7 + 0.3\exp(-t/1000). \tag{8.4}$$

where t is the number of days from injection into GEO. We are now able to write:

$$P = P_0 J D \cos\theta_s' \tag{8.5}$$

and insert values for J, D and θ_s' from Eqns 8.2 through 8.4. Figure 8.15 shows $P = f(t)$ with $t = 0$ on 1 January such that $DD \simeq t$.

A positive power balance exists as long as the load profile remains below the undulating array profile. It may happen towards the end of a satellite's lifetime, however, that a negative balance occurs for a few minutes or hours per day, depending on the switching schedule of spacecraft equipment. The deficit is most likely to occur around the solstices, where the solar array output is at its lowest, as the solar incidence angle approaches 23.4°. If nothing is done, the battery will

Fig. 8.15. Typical solar array power output profile over 7 years. A, summer solstice; B, autumn equinox; C, winter solstice; D, vernal equinox.

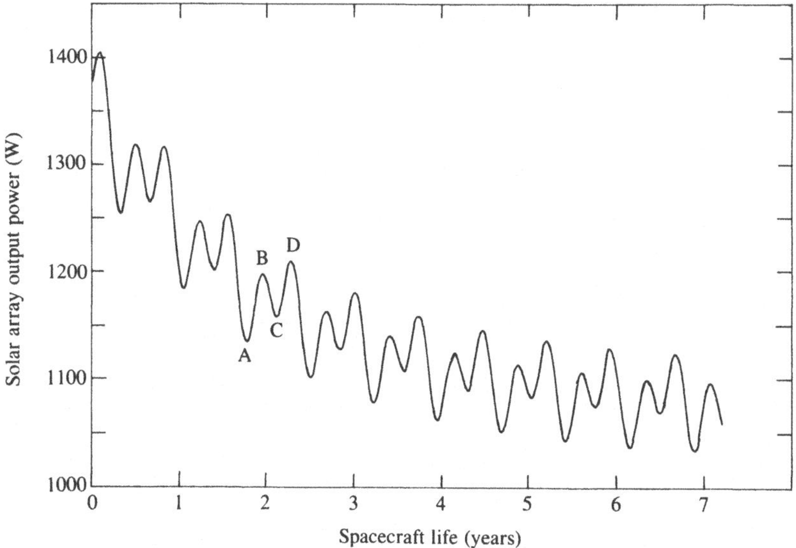

begin to discharge to make up the deficit. This may be acceptable for a while but, as the discharge durations become longer, one would have to consider switching off non-essential loads such as heaters and amplifiers, even if such actions were to degrade the value of the mission.

Bibliography

Photovoltaic Generators in Space (1982). ESA SP-173. Paris: European Space Agency.

Power Electronics Specialists Conference (1985). ESA SP-230. Paris: European Space Agency.

Solar Cell Array Design Handbook (1976). Vols 1 and 2. Pasadena: Jet Propulsion Laboratory.

9

Propulsion and Orbit Control

Introduction

In Chapter 1 we looked at propulsion and orbit acquisition in the context of launch vehicles. We established that only one vehicle (Proton) is capable of routinely injecting satellite payloads directly into geostationary orbit, and that passengers onboard any of the other vehicles had to be equipped with rocket engines of their own to accomplish the final leg of the voyage. In Chapter 3 we examined the causes of north–south and east–west drift of a geostationary satellite around its nominal sub-satellite point. We also suggested strategies for drift correction using thrusters. The design and use of these onboard propulsion devices is the subject of the present chapter.

Propulsion

Bipropellant Subsystem Architecture

Figure 9.1 shows the design of a modern bipropellant propulsion subsystem. The MMH fuel and the N_2O_4 oxidizer are stored in separate tanks. Pipes run from the tanks to a large thruster and to 12 smaller thrusters. Each thruster has a valve to admit propellants. When the fuel and the oxidizer mix in the thrust chambers, they ignite spontaneously. The combustion generates thrust forces. The large thruster is an apogee kick motor (AKM), while the smaller thrusters are used for orbit and attitude control.

The round tank at the top of the figure contains helium gas to pressurize the fuel and oxidizer tanks. Pressurization fulfils the same function as turbopumps in launch vehicle engines, namely to feed propellant to the thrusters at a controlled rate. Each tank is equipped with a *pressure transducer* (PT) for performance monitoring via telemetry.

Fig. 9.1. Typical bipropellant propulsion subsystem architecture. Courtesy of British Aerospace and INMARSAT.

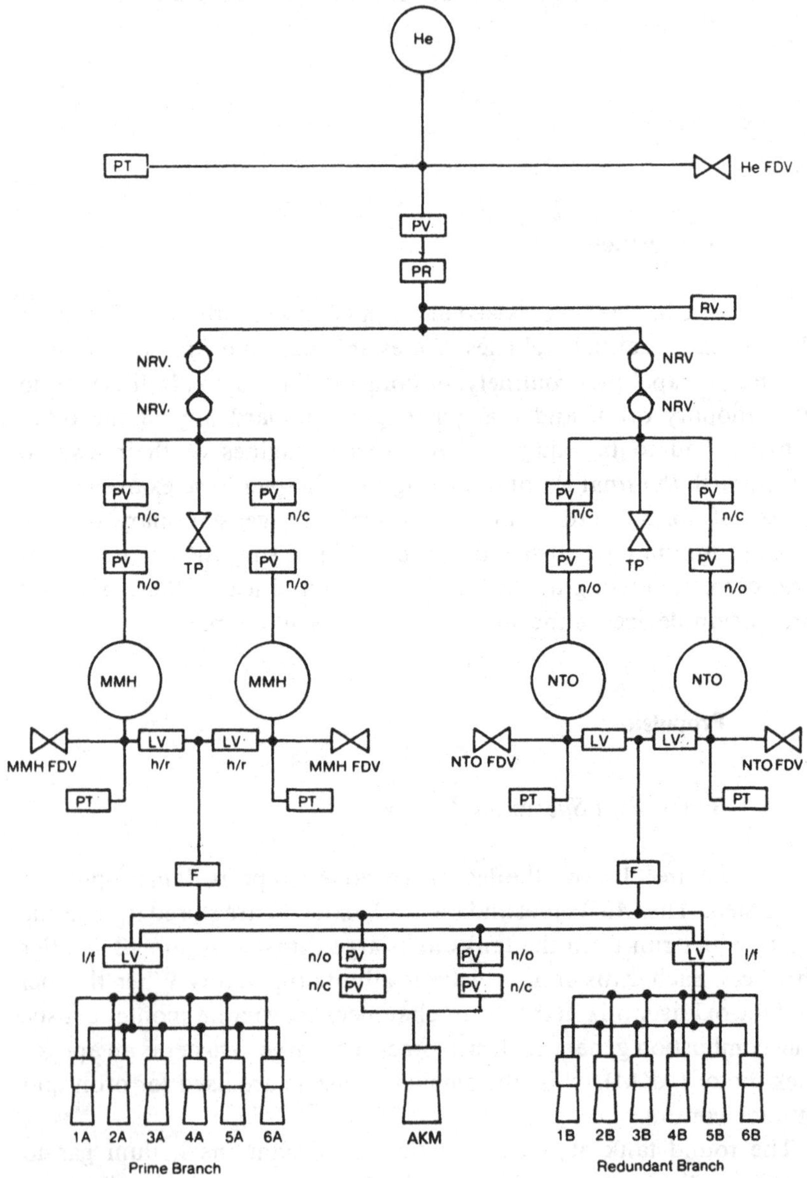

Non-return valves (NRV) and *pyrotechnic valves* (PV) are installed between the pressurant tank and the propellant tanks. Non-return valves prevent accidental mixing of fuel and oxidizer through the pressurizing pipe. Pyro valves are actuated by explosives and can only be opened or closed once. They are lighter and more reliable than other valve types, although their once-only mode of operation limits their usefulness. The valves marked "n/o" are "normally open", i.e. they are open when installed and close for good once their pyrotechnic actuators are powered by telecommand. Similarly, the "n/c" valves are "normally closed" and may be opened once and for all by telecommand.

The n/c valves are opened moments before the AKM firing to ensure adequate propellant flow. When the apogee kick operation is completed, the n/o valves are closed, and the pressurant tank is thus isolated for the remainder of the mission.

The reason for isolating the pressurant tank is that the small thrusters require lower propellant pressure than the AKM. The small thrusters are fed from the propellant tanks in a *blow-down mode*, i.e. the residual tank pressure is adequate for their operation even when the pressure decreases as propellant is consumed.

The n/o pyro valves in the pipes feeding the AKM are telecommanded to close once the motor is no longer needed. Isolating the AKM is a precautionary measure to avoid propellant leakage later in life through the motor's thrust chamber valves.

Electromagnetic *latch valves* (LV) may be opened and closed at will and are used instead of pyro valves in the thruster feed lines. These valves are opened by telecommand before an attitude or orbit manoeuvre, and are closed after the manoeuvre, to prevent leakage through the thrusters when not in use.

The *filters* prevent stray solid particles from becoming stuck in valve seats and causing leakage. The particles could have been left behind inadvertently after subsystem manufacturing and cleaning, or they may have been accidentally introduced into the system during the pre-launch propellant filling process.

Each of the tanks is provided with a *fill-and-drain valve* (FDV) which allows propellant to be loaded and unloaded during ground test prior to launch.

Monopropellant Subsystem Architecture

Monopropellant propulsion subsystems, such as those employing hydrazine, only require a single propellant tank, although two or more tanks are frequently installed for balancing purposes. In spin-stabilized satellites, the centrifugal force and the gas acting on the propellant is sufficient for feeding the thrusters, and no pressurant tank is needed. The number of valves may therefore be reduced, which saves mass and increases reliability. Three-axis-stabilized satellites require either a pressurant or an internal tank *diaphragm* ("bladder") to force the monopropellant to the thrusters. Diaphragms are simple and reliable, but because their expulsion efficiency is inferior to pressurized systems, propellant is wasted due to the relatively large quantity of residuals remaining in the tank at the end of spacecraft life. Diaphragms are not suitable for bipropellant propulsion subsystems due to their inability to withstand the highly corrosive properties of oxidants.

Combustion of monopropellant hydrazine occurs as it passes across a *catalyst bed* inside a thruster (Fig. 9.2). The catalyst usually consists of iridium deposited on porous ceramic pellets made of aluminium oxide. Hydrazine decomposes into ammonia, hydrogen and nitrogen. In some satellites the hot combustion gases are further heated by electrical means to improve the specific impulse I_{sp}.

Liquid monopropellants are adequate for small thrusters, but less so for AKM use where the propellant flow rate is much higher. Monopropellant satellites therefore employ *solid propellant* AKMs which are functionally independent of the liquid subsystem.

Architectural Variations

Some satellites use a combination of mono- and bipropellant propulsion methods; others make both the perigee and the apogee motor functions an integral part of the liquid propulsion subsystem design, thereby obviating the need for a separate PKM upper stage.

Fig. 9.2. Monopropellant catalytic thruster with heaters.

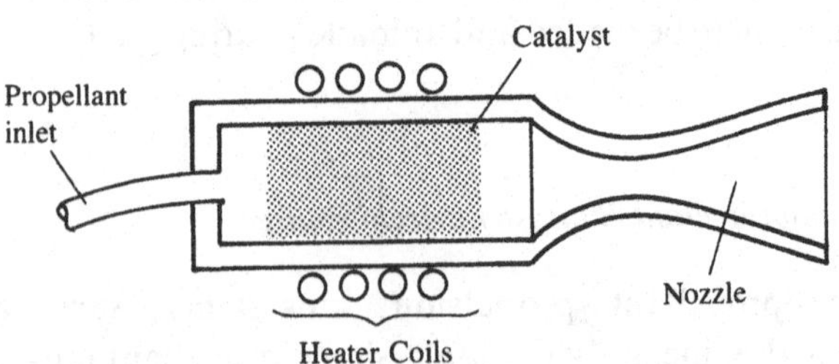

Orbit Control

Perigee and apogee kick motors perform orbit control functions in parking and transfer orbits, and thrusters do the same in geostationary orbit. In this section we will study methods of carrying out orbit control. We will also calculate the amount of propellant mass needed for the various stages of control. To accomplish this, we need to recall Eqns 1.3 and 1.6 from Chapter 1. Thus:

$$\Delta V = I_{sp} g \ln (m_1/m_2) \tag{9.1}$$

$$\Delta V = (V_1^2 + V_2^2 - 2V_1V_2 \cos \Delta i)^{1/2}. \tag{9.2}$$

GTO–GEO Manoeuvre Strategy

The apogee kick motor performs three different manoeuvres at the end of the transfer orbit phase: (i) circularization, (ii) inclination change and (iii) node rotation. In theory, the three manoeuvres could be combined in a single motor burn. In the practical case of a transition from GTO to GEO, only circularization and inclination control can be achieved simultaneously, and only if the GTO apogee coincides with an orbital node (i.e. $\omega = 0°$ or $180°$). The caveat arises because circularization to GEO must occur at apogee, and because inclination reduction without node rotation can only be achieved at the nodes (Fig. 9.3). The

Fig. 9.3. Inclination change without node rotation.

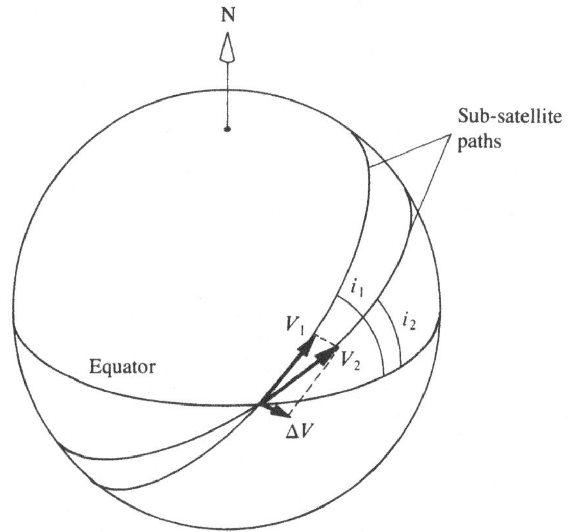

reason why most expendable launch vehicles have a coasting phase before last stage ignition is precisely to allow the perigee and the apogee to land above the equator, i.e. at the nodes.

Node rotation, on the other hand, requires an impulse away from the nodes (Fig. 9.4). It is therefore necessary to perform the manoeuvre

Fig. 9.4. Combined inclination change and node rotation.

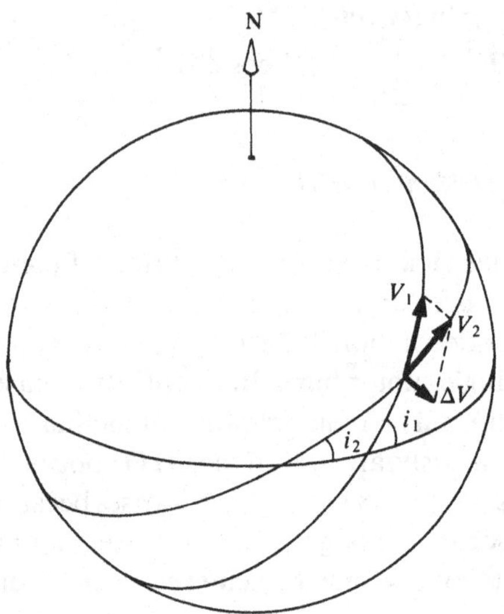

Fig. 9.5. Definition of angles.

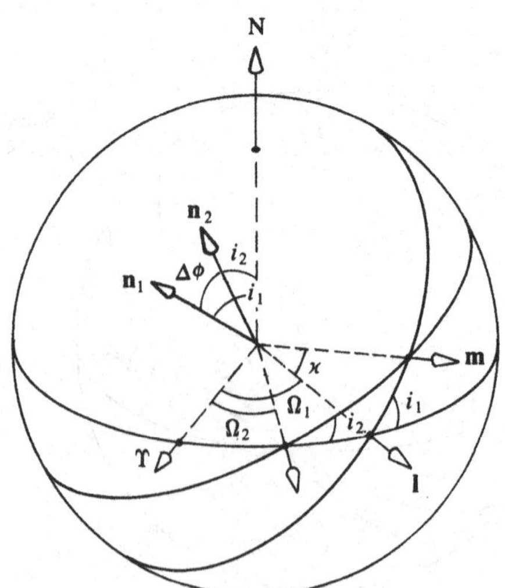

either before or after AKM burn. Normally it is done after the burn, for node rotation consumes less propellant when the inclination is low.

The above discussion is particularly valid for a *solid* propellant AKM. *Liquid* propellant AKMs are restartable, and the transition from GTO to GEO may be conducted in two or more steps. Greater flexibility therefore exists to combine manoeuvres and hence to optimize fuel consumption.

If we denote GTO elements with the subscript "1" and the GEO elements with "2", then the orbit plane normal vectors are found to be (Fig. 9.5):

$$\mathbf{n}_1 = (\cos\Omega_1 \sin i_1, -\sin\Omega_1 \sin i_1, \cos i_1),$$

$$\mathbf{n}_2 = (\cos\Omega_2 \sin i_2, -\sin\Omega_2 \sin i_2, \cos i_2).$$

The orbit plane change $\Delta\phi$ becomes:

$$\cos\Delta\phi = \mathbf{n}_1 \cdot \mathbf{n}_2 \tag{9.3}$$

Note that $\Delta\phi \neq \Delta i$, unless $\Delta\Omega = 0°$.

Fig. 9.6. Definition of angles.

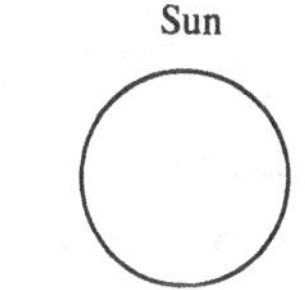

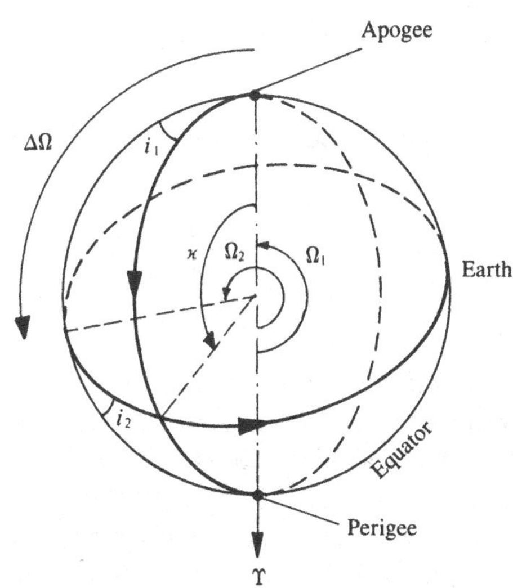

To find out how far from the ascending node the AKM should be fired, define κ as shown in Fig. 9.5. Then:

$$\mathbf{l} = (\cos \Omega_1, \sin \Omega_1, 0),$$
$$\mathbf{m} = (\mathbf{n}_2 \times \mathbf{n}_1)/\sin \Delta\phi,$$
$$\cos \kappa = \mathbf{l} \cdot \mathbf{m} = \mathbf{l} \cdot (\mathbf{n}_2 \times \mathbf{n}_1)/\sin \Delta\phi. \tag{9.4}$$

For example, take a satellite with a lift-off mass of 1300 kg launched from Cape Canaveral (GTO inclination = 28°) into a transfer orbit with r_p = 6578 km and r_a = 42 164 km. Equation 2.3 gives us a_1 = 24 371 km and Eqn 2.9 yields V_1 = 1.597 km/s at apogee. Lift-off is at 23.30 GMT on 21 September. Thirty minutes later, at 00.00 GMT, the satellite is injected at perigee over the equator. The argument of perigee ω_1 is therefore 180°. The descending node happens to coincide with the Greenwich meridian. According to Eqn 2.16 in Chapter 2, Ω_1 = 180°, which is confirmed by studying Fig. 9.6. We have thus defined the GTO in terms of i_1, Ω_1, ω_1 and V_1.

For our propellant mass calculations, assume that the satellite AKM and thrusters have a specific impulse I_{sp} of 290 s.

We now wish to achieve a GEO with i_2 = 3° and Ω_2 = 280° so that the inclination vector drifts through 0° after a few years (Fig. 3.5). Here a_2 = 42 164 km and V_2 = 3.075 km/s.

The first step is to circularize the GTO and lower the inclination from 28° to 3° by firing the AKM at apogee. $\Omega_1 = \Omega_2$ = 180° for the time being. Equation 9.3 gives us:

$$\cos \Delta\phi = \cos(28° - 3°); \qquad \Delta\phi = \Delta i = 25°$$

and Eqn 9.2 leads to:

$$(\Delta V)^2 = 1.597^2 + 3.075^2 - 2 \times 1.597 \times 3.075 \cos 25°,$$
$$\Delta V = 1.762 \text{ km/s}.$$

The propellant mass Δm consumed to produce this ΔV is obtained from Eqn 9.1:

$$\Delta m = m_1 - m_2 = m_1 \left\{1 - \exp\left(\frac{-\Delta V}{g I_{sp}}\right)\right\}, \tag{9.5}$$

$$\Delta m = 600 \text{ kg},$$

and the new satellite mass m_1 becomes 1300 − 600 = 700 kg. In other words, moving from GTO to GEO consumes nearly 50% of the satellite lift-off mass, or 80–90% of all the onboard propellant.

If no inclination change had been performed (Δi = 0), then ΔV would have reduced to 1.478 km/s for circularization only. A subsequent inclination change of the GEO would have required an additional ΔV

= 1.331 km/s. In total, such a sequential procedure would have demanded 1.478 + 1.331 = 2.809 km/s, which is far greater than the 1.762 km/s needed for a combined manoeuvre.

The next step is to rotate the GEO nodes by $\Delta\Omega = 280 - 180 = 100°$ while keeping i constant at 3°. From Eqn 9.3:

$$\cos \Delta\phi = \sin^2 i \cos(\Omega_2 - \Omega_1) + \cos^2 i,$$
$$\Delta\phi = 4.6°.$$

Equation 9.2 gives:

$$(\Delta V)^2 = 2 \times 3.075^2(1 - \cos 4.6°),$$
$$\Delta V = 0.247 \text{ km/s} = 247 \text{ m/s},$$

and from Eqn 9.4:

$$\kappa = 130°.$$

We found earlier than the satellite mass was reduced from 1300 to 700 kg during the circularization and inclination change manoeuvres. Equation 9.5 gives us the propellant mass consumed during the node rotation manoeuvre:

$$\Delta m = 700\left\{1 - \exp\left(\frac{-247}{9.81 \times 290}\right)\right\} = 58 \text{ kg}.$$

The total cost in propellant mass to take our particular satellite from GTO to the final GEO is therefore:

$$\Delta m = 600 + 58 = 658 \text{ kg}.$$

The remaining satellite mass is 1300 − 658 = 642 kg in our example.

The orbital manoeuvre strategy described in this example is not necessarily the most efficient in terms of propellant consumption, but is offered merely as an illustration of the principles involved. Mission analysts have developed propellant management to a fine art and often get into heated debates about the best approach to a particular mission.

Geostationary Orbit Control

In Chapter 3 we studied how gravitational forces from the moon and the sun give rise to orbital inclination drift (called north–south drift), while irregularities in the earth's gravitational field cause geostationary satellites to wander in an east–west direction. Propellant is needed to correct these drift tendencies for the duration of the satellite mission.

North–South Station-Keeping

North–south station-keeping is synonymous with *inclination control* or *inclination correction*, i.e. the art of performing manoeuvres to contain the inclination within specified limits. Most geostationary satellites are maintained inside an *inclination window* for as long as there is propellant left onboard. The upper limit of the window is typically ±0.1° for a communications satellite equipped with antenna spot beams. Maintaining a satellite within such a tight boundary requires frequent inclination control, perhaps once every 3 months. A scientific satellite may have more relaxed constraints on the window – say, 3°. In this case, inclination control would rarely be needed, if ever, during the satellite's orbital lifetime.

As explained in Chapter 3, drift movement of the inclination vector translates into drift of i and Ω (Fig. 3.5). A 3° inclination window has been plotted in Fig. 9.7 along with a drift trajectory which maintains the inclination within the window during 7.5 years. As the vector is about to exceed the 3° limit (point B), it may be desirable to bring it all the way back to the start of the trajectory (A), so as to avoid further inclination corrections during the following 7.5 years.

Let us calculate ΔV and Δm to move from B to A. With m_1 = 642 kg, $\Omega_1 = 80°$, $\Omega_2 = 280°$, $i_1 = i_2 = 3°$, we obtain from Eqns 9.3 and 9.2:

$$\Delta V = 316 \text{ m/s}$$

and from Eqn 9.5:

Fig. 9.7. Movement of the inclination vector within a 3° inclination window.

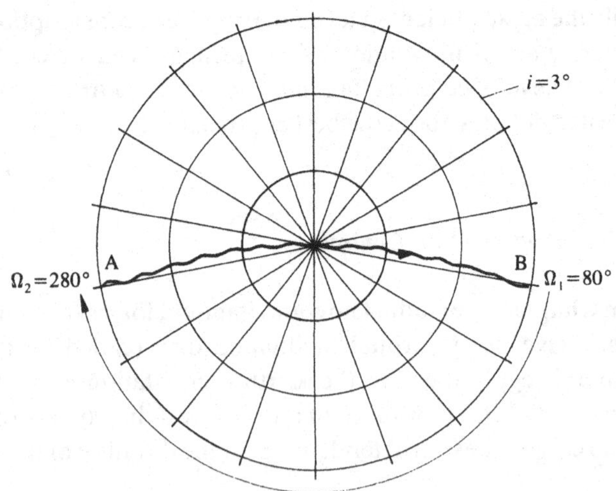

$$\Delta m = 642\left\{1 - \exp\left(\frac{-316}{9.81 \times 290}\right)\right\} = 68 \text{ kg}.$$

Geostationary satellites equipped with narrow-beam antennae (Chapter 12) are maintained within an inclination window of 0.1° to avoid the antenna footprint on the earth meandering due to the satellite's diurnal "figure-8" movement. Given that $V_1 = V_2 = V = 3075$ m/s, $i_1 = i_2 = i$, and $\Delta\Omega = 180°$, the required ΔV for north–south station-keeping is obtained from Eqns 9.2 and 9.3:

$$\Delta V = 6148 \sin i.$$

Inserting $i = 0.1°$ gives;

$$\Delta V = 10.73 \qquad (\text{m/s}).$$

The propellant mass consumption depends on the satellite mass and the I_{sp} of the thrusters used, according to Eqn 9.5. Since the inclination drift rate is approximately 0.8° per year (Eqn 3.2), a north–south manoeuvre is required every 3 months. Figure 9.8 shows the amount of propellant needed for controlling the inclination of a 500-kg satellite. The propellant mass consumption is a linear function of the satellite mass according to Eqn 9.5, so the consumption for other satellite masses is easy to calculate by extrapolation.

Fig. 9.8. Propellant mass consumption for a north–south station-keeping manoeuvre of a 500-kg satellite.

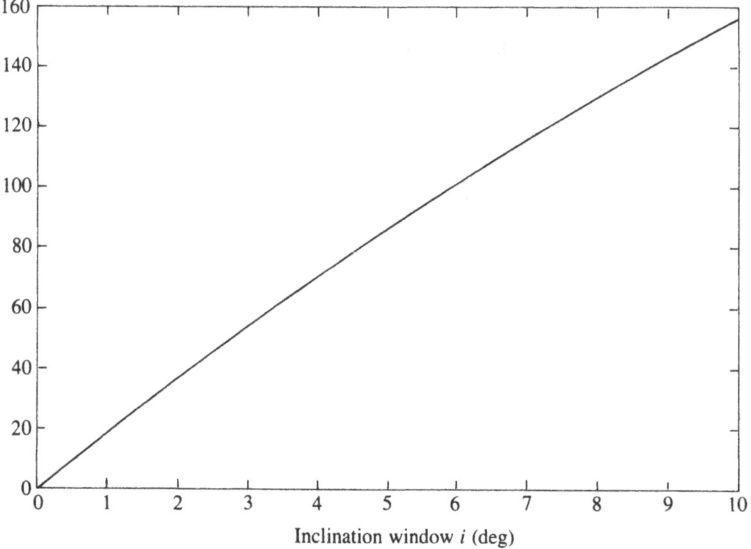

East–West Station-keeping

East–west station-keeping means *longitude control* or *longitude correction* of the sub-satellite point. In analogy with north–south station-keeping, one defines a *longitude window*. According to recommendations of the International Telecommunications Union (ITU), geostationary satellites should be maintained within a longitude window of $\pm 0.1°$ to avoid mutual radio interference with neighbouring satellites. The total width of the window is therefore 0.2°.

The most common strategy for east–west control consists in allowing the satellite to drift to the specified longitude limit, and then giving it an impulse in the opposite direction such that it touches the other limit before returning (Fig. 9.9). The movement resembles that of a tennis ball being bounced vertically off a racket. As soon as the ball returns to the racket, it is given an impulse to repeat the movement.

From Chapter 3 we recall that a geostationary satellite will tend to drift along the equator towards the nearest stable point (Fig. 3.9). Equation 3.3 gave the approximate longitude drift acceleration rate as:

$$\ddot{\Lambda} = C \sin(150° - 2\Lambda) \qquad (\text{deg/day}^2) \tag{9.6}$$

where

$$C = \{16.9 + 2.9 \sin(\Lambda - 35°)\} \times 10^{-4}.$$

Let $\Delta\Lambda$ be the half-width of the drift window, e.g. $\Delta\Lambda = 0.1°$. The ΔV required for a correction manoeuvre is:

Fig. 9.9. Satellite drift velocity inside an east–west station-keeping window.

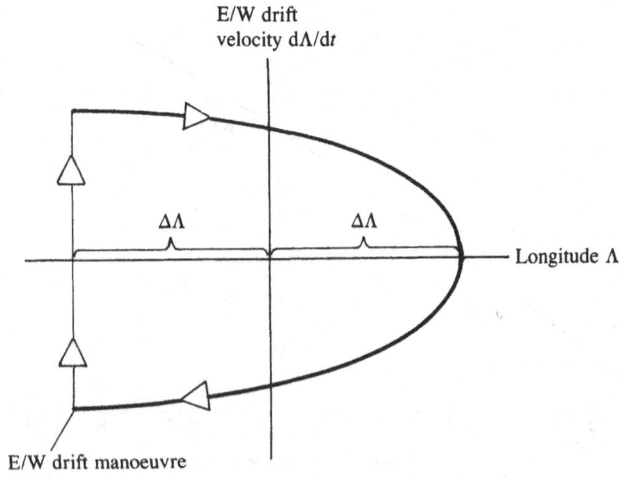

$$\Delta V = 11.36\sqrt{\ddot{\Lambda}\Delta\Lambda} \qquad \text{(m/s)}. \tag{9.7}$$

The period τ of the satellite pendulum movement through the window is:

$$\tau = 4\sqrt{\Delta\Lambda/\ddot{\Lambda}} \qquad \text{(days)} \tag{9.8}$$

and, consequently:

$$\Delta V = 45.44\, \Delta\Lambda/\tau \qquad \text{(m/s)}. \tag{9.9}$$

As before, $\Delta\Lambda$ is in degrees and τ in days.

Equations 9.7 and 9.8 have been plotted in Figs 9.10 and 9.11 for $\Delta\Lambda = 0.1°$ and for various values of the nominal station-keeping longitude Λ. The corresponding propellant mass consumption for several satellite masses is shown in Fig. 9.12, whereby:

$$\frac{\Delta m}{\text{year}} = \frac{\Delta m}{\text{manoeuvre}} \cdot \frac{360°}{\tau}.$$

It can be seen that east–west station-keeping consumes far less propellant than north–south corrections, but manoeuvres must be performed quite frequently (as often as once per month) except in the vicinity of the stable and unstable points. The total amount of propellant required for east–west station-keeping is, in the long term, independent of the frequency of corrections, and depends solely on the nominal longitude.

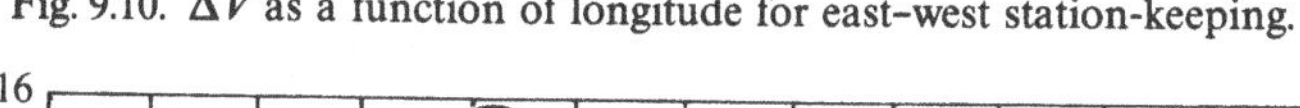

Fig. 9.10. ΔV as a function of longitude for east–west station-keeping.

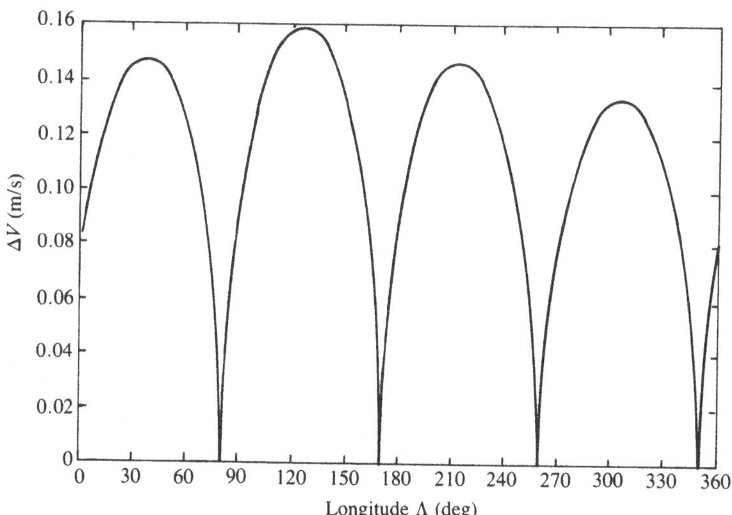

Satellite Repositioning

Sometimes it is desirable to move a satellite from one sub-satellite point to another. The move is called *satellite repositioning*. The

Fig. 9.11. Period between east–west station-keeping manoeuvres.

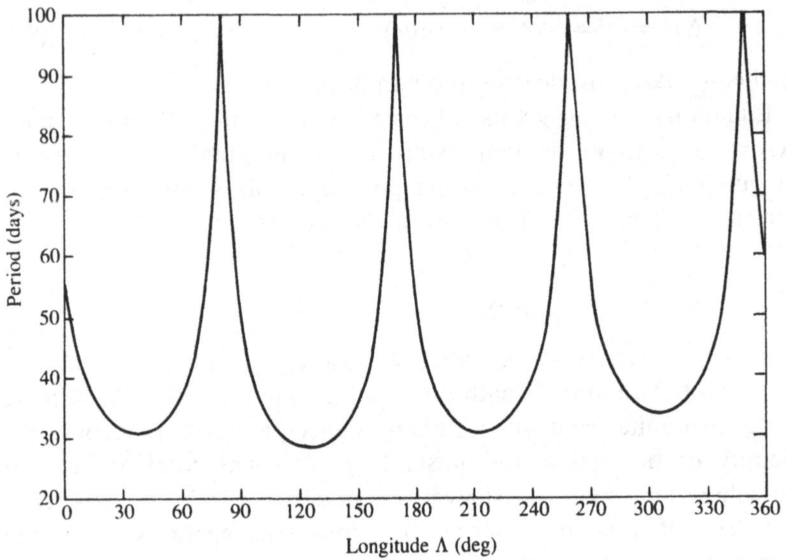

Fig. 9.12. Propellant mass consumption, for 1 year of east–west station-keeping of a 500-kg satellite.

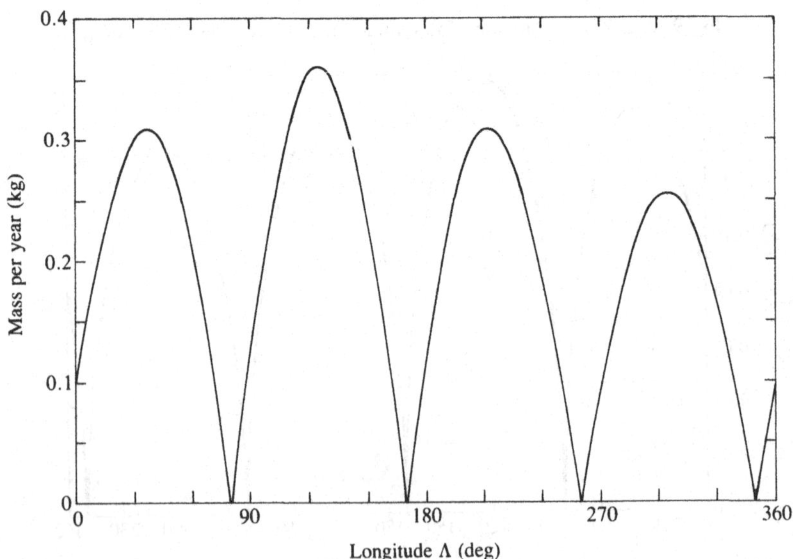

Table 9.1. *Propellant mass budget for a 1300-kg satellite maintained for 10 years within 0.1° in E/W and 3° in N/S*

	ΔV (m/s)	$1 - e^{-\Delta V/g \cdot I_{sp}}$	m_i (kg)	Δm (kg)
Circularization and inclin. reduction	1762	0.46171	1300	600
Node rotation	247	0.083	700	58
GTO Attitude ctrl	—	—	642	2
1st year E/W	0.159	0.000054	640	0.4
1st year att. ctrl	—	—	639.6	0.1
2nd year E/W	0.159	0.000054	639.5	0.4
2nd year att. ctrl	—	—	639.1	0.1
3rd year E/W	0.159	0.000054	639	0.4
3rd year att. ctrl	—	—	638.6	0.1
Relocation from 120°E to 210°E at 3° per day	17	0.00596	638.5	3.8
4th year E/W	0.148	0.000052	634.7	0.4
4th year att. ctrl	—	—	634.3	0.1
5th year E/W	0.148	0.000052	634.2	0.4
5th year att. ctrl	—	—	633.8	0.1
6th year E/W	0.148	0.000052	633.7	0.4
6th year att. ctrl	—	—	633.3	0.1
7th year E/W	0.148	0.000052	633.2	0.4
7th year att. ctrl	—	—	632.8	0.1
N/S ctrl at $i = 3°$ by node rot. 180°	322	0.107	632.7	67.7
8th year E/W	0.148	0.000052	565	0.4
8th year att. ctrl	—	—	564.6	0.1
9th year E/W	0.148	0.000052	564.5	0.4
9th year att. ctrl	—	—	564.1	0.1
10th year E/W	0.148	0.000052	564	0.4
10th year att. ctrl	—	—	563.6	0.1
Graveyard to a = GEO + 200 km	6	0.00211	563.5	1.2
Total/final	2356		562.3	737.7

associated ΔV is not a function of the distance, but merely of the average longitude drift rate.

The ΔV needed to obtain a certain longitude velocity is:

$$\Delta V = 2.83 \dot{\Lambda} \quad \text{(m/s)} \tag{9.10}$$

where $\dot{\Lambda}$ is the drift rate in degrees/day. For example, a ΔV of 8.5 m/s is required to obtain a longitude drift rate of 3° per day, and the same ΔV

is needed to stop the drift once the destination has been reached. The total ΔV is therefore 17 m/s.

Equation 9.7 is only valid for small $\Delta\Lambda$. In reality, any ΔV has the effect of changing a as well as e according to Eqns 2.10 and 2.5, assuming that r is constant for the duration of thruster firing. Because the orbit is no longer truly circular, a satellite being repositioned will experience a 24-h velocity modulation around the mean drift rate as it passes through the newly formed apogee and perigee.

A *graveyard burn* is a special form of repositioning where a satellite is removed from the geostationary orbit once and for all when the spacecraft is considered unusable. The manoeuvre is performed in two steps. A thruster is fired to create an apogee beyond geosynchronous altitude. A burn at apogee follows so as to raise the perigee to the same height. Graveyard burns typically require a $\Delta V = 7$ m/s.

Propellant Mass Budget

Table 9.1 shows a step-by-step mass budget based on similar data as in the above analysis. The interested reader might study whether the total propellant consumption changes if the position of the north–south station-keeping manoeuvre in the chronology is altered.

Bibliography

Soop, E. M. (1983). *Introduction to Geostationary Orbits*, ESA SP-1053. Paris: European Space Agency.

Sutton, G. P. (1986). *Rocket Propulsion Elements*. New York: John Wiley.

10

Attitude Stabilization, Measurement and Control

Introduction

A geostationary applications satellite must be orientated in space such that its antennae or radiometers view the earth continuously. The solar arrays should also face the general direction of the sun at all times. Given that the satellite-earth and the satellite-sun vectors move 360° relative to each other every day, the satellite has to be something of a contortionist to satisfy both pointing conditions.

The obvious design solution to the variable two-way pointing problem is to mount the antennae or telescopes on one part of the spacecraft and the solar arrays on another, allowing the two parts to rotate in opposite directions around a common shaft. Two-way pointing can be maintained as long as the orientation or *attitude* of the shaft remains approximately perpendicular to the earth and sun vectors. Hence the dual spin and the three-axis stabilized design concepts (Figs 10.1–10.4).

The attitude of a spacecraft is the orientation of its body axes in inertial space. The angles which define an attitude may be direction cosines or azimuth and elevation (Fig. 10.5); the latter convention was used in Chapter 2 to derive an expression for sun angle.

Although space is virtually void of matter, it is nevertheless full of forces acting on the spacecraft (see Chapter 4), and some of these cause the attitude to drift. If the two-way pointing requirement is to be met at all times, it is necessary to ensure attitude *stabilization* continuously, and perform *measurement* and *control* at regular intervals. The present chapter describes how this is done.

Fig. 10.1 (Left). Satellite spin stabilization concept. Fig. 10.2 (Right). Satellite dual spin stabilization concept.

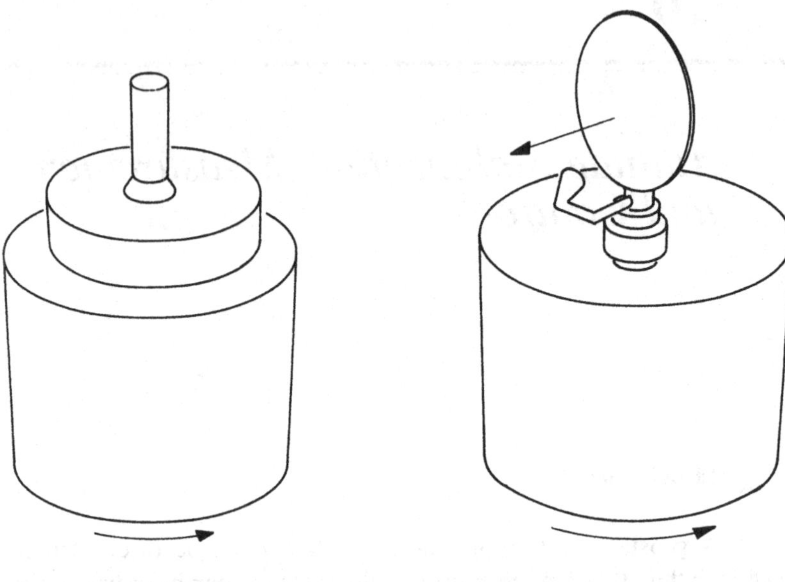

Fig. 10.3 (Left). Satellite three-axis stabilization concept using a bias momentum wheel. Fig. 10.4 (Right). Satellite three-axis stabilization concept using reaction wheels.

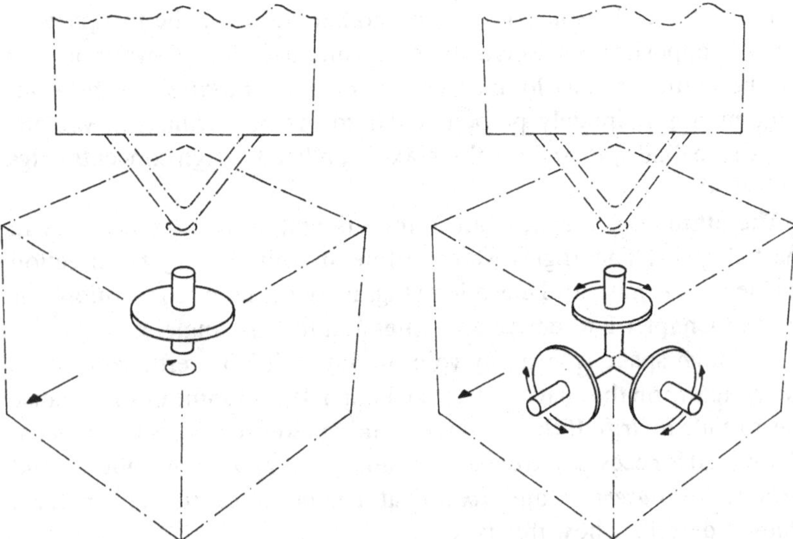

Gyroscopic Theory

The *gyroscopic effect* is one of nature's most bizarre gifts to mankind. If an object in space is rotated, it will continue to spin around a momentum vector whose direction remains fixed in inertial space. The attitude of the object will drift if perturbed by external or internal torques, but the perturbation will be resisted to a degree which is proportional to the spin rate. This resistance is often referred to as *gyroscopic stiffness.*

In spite of its resistance to torques, a gyroscope will nevertheless yield and, in doing so, move in an unexpected direction. Apply a torque $\vec{T}$ through the centre of gravity at 90° to the momentum vector $\vec{H}$. One would intuitively expect the spin axis to rotate about $\vec{T}$ (Fig. 10.6) with an angular acceleration $\dot{\Omega} = T/I$. In reality, the spin axis vector tries to align itself with the torque vector $\vec{T}$. If the torque vector "haunts" the satellite such that it remains perpendicular to the spin axis while the latter is moving, the axis will precess with a constant angular speed $\Omega = T/H$ (Fig. 10.7).

The reason for this surprising behaviour of the spin axis is worth exploring. Let the disc in Fig. 10.8 be a cross-section of a cylindrical body. The disc is rotating at a rate ω and is also free to rotate about

Fig. 10.5. Definition of spacecraft attitude angles. See also Fig. 2.15.

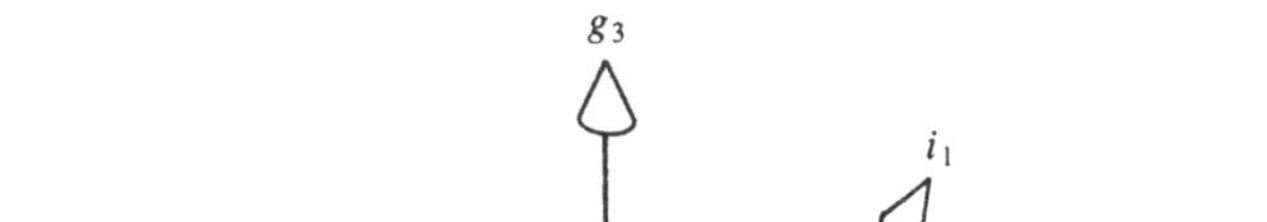

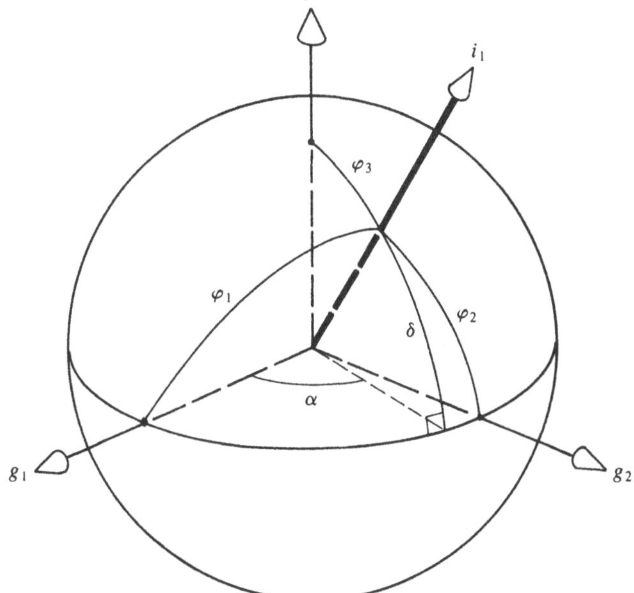

Fig. 10.6. Intuitive attitude response to external torque $\vec{T}$.

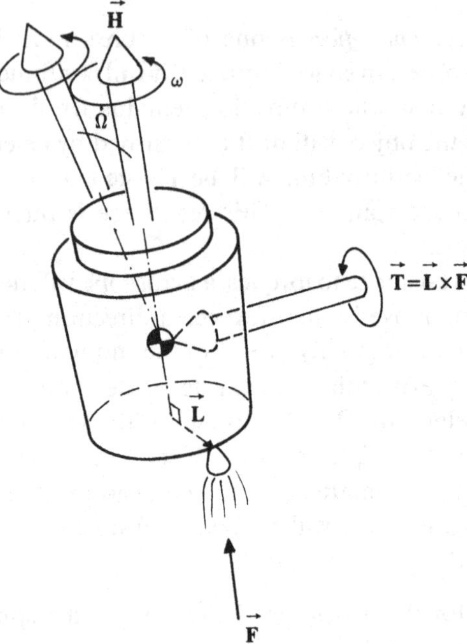

Fig. 10.7. Actual attitude response to external torque $\vec{T}$ due to gyroscopic effect.

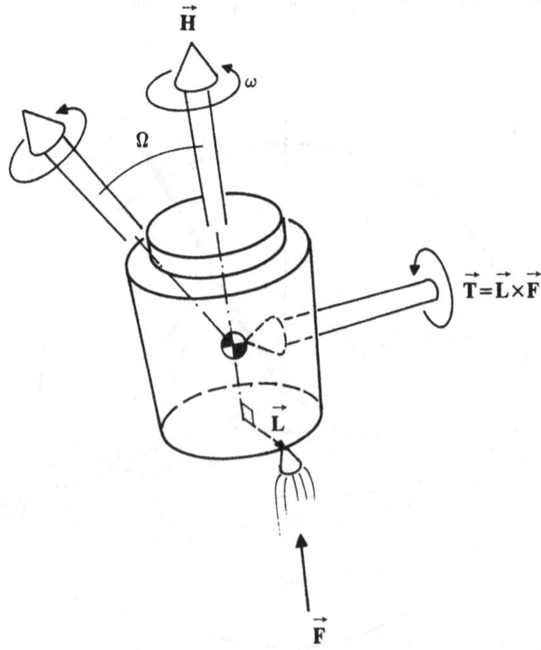

XX' and *YY'*. Now let *dm* represent a mass element of the disc in sector *A*, and apply a torque $\vec{\mathbf{T}}$ along *XX'*. The torque will impart an instantaneous angular acceleration $\ddot{\varphi}$ to *dm* about *XX'*. The acceleration is enhanced by the motion of *dm* as it approaches *XX'*, since the moment of inertia I_{XX} due to *dm* decreases. The acceleration increment is represented in Fig. 10.8 by an inertial force F_A pointing *towards* the reader and applying at *dm*.

As the mass element moves through sector *B*, the angular acceleration $\ddot{\varphi}$ is counteracted as *dm* moves away from *XX'* and I_{XX} increases. The counteraction may be represented by an interial force F_B which again points *towards* the reader.

By following the path of *dm* through sectors *C* and *D*, we find by similar reasoning that the equivalent inertial forces F_C and F_D point *away* from the reader. It follows that $F_A + F_B$ and $F_C + F_D$ form a couple which causes the disc to rotate about *YY'*. Moreover, with $F_A = F_B$ and $F_C = F_D$, there is no average rotation around *XX'*.

The forces acting on our chosen *dm* will of course also act on any other mass element in the body, such that the entire body is subjected to the gyroscopic effect.

Our planet earth behaves like a gyroscope. It turns around its axis at a rate of 360° in 23 h and 56 min, its momentum axis tilted 23.44° to the ecliptic plane normal. In the short term, the attitude of the earth's spin axis is fixed in inertial space, but in the course of 26 000 years it actually precesses one complete turn due to the gravitational pull of the sun and the moon on the earth's equatorial bulge.

If a perturbing torque $\vec{\mathbf{T}}$ is applied to a spinning object for a duration which differs from the spin period, the object will be thrown off balance and be left in a state of *nutation*. To illustrate this phenomenon, take a rigid, axisymmetric cylinder, i.e. one whose lateral moments of inertia are equal due to symmetry ($I_x = I_y$) and whose longitudinal moment of inertia axis I_z constitutes the spin axis **z**. The attitude of the body-fixed momentum vector $\vec{\mathbf{H}}$ has drifted some angular distance θ from the system momentum vector $\vec{\mathbf{M}}$ due to $\vec{\mathbf{T}}$, and the spin axis **z** is now precessing around $\vec{\mathbf{M}}$ at an angle θ and at a rate $\dot{\psi}$ (Fig. 10.9). We have:

$$\dot{\psi} = \frac{I_z}{(I_x - I_z)\cos\theta}\,\dot{\eta} \tag{10.1}$$

with $\dot{\eta}$ being the spin rate in the body-fixed reference frame. The angular rates $\dot{\psi}$ and $\dot{\eta}$ are in fact components of the pre-nutation satellite spin rate ω_z:

$$\omega_z = \dot{\eta} + \dot{\psi} \cos \theta. \tag{10.2}$$

After combining Eqn 10.2 with Eqn 10.1 and eliminating $\dot{\eta}$, we obtain the nutation rate as:

$$\dot{\psi} = \frac{I_z}{I_x \cos \theta} \omega_z \tag{10.3}$$

By examining the sign of Eqn 10.1, we find that the spin axis nutation is *prograde* for a tall, thin-cylinder such as a rocket stage where $I_z < I_x$. Conversely, the nutation is *retrograde* for a large, flat disc with $I_z > I_x$.

Fig. 10.8. The gyroscopic effect.

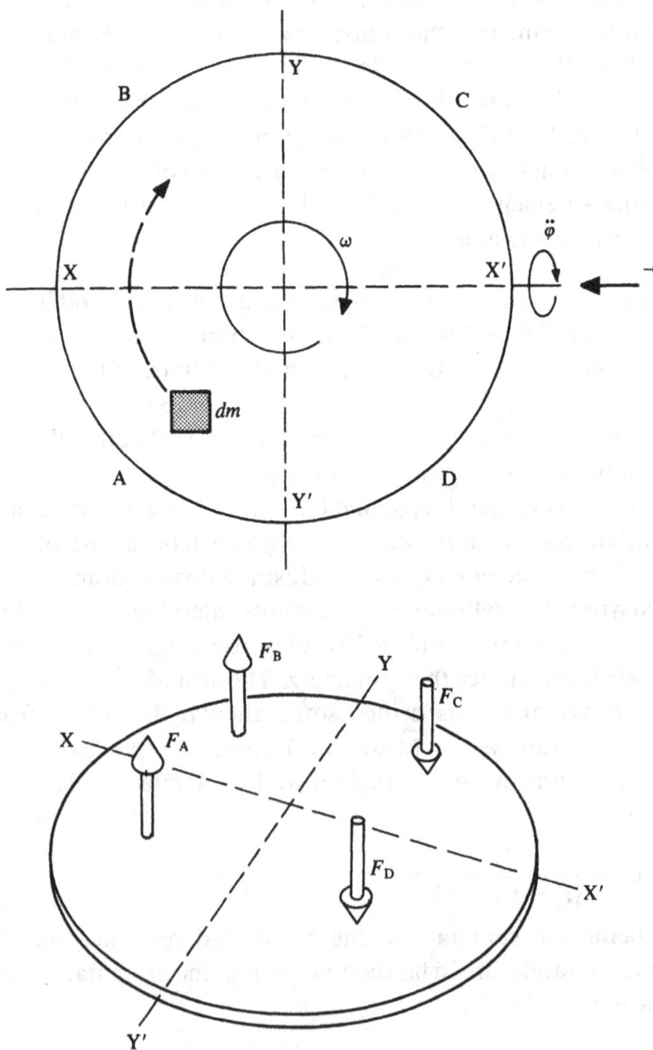

Internal dissipation of angular momentum will cause the body to re-align itself with the original spin axis. The rate of dissipation determines the speed with which alignment is achieved. This phenomenon is exploited to eliminate nutation through *nutation damping* (see below).

Passive Stabilization

From the above discussion it is obvious that the simplest method of stabilizing a spacecraft is to make it spin around its major primary axis of inertia. The method is called *spin stabilization* (Fig. 10.1) and relies on the gyroscopic effect to establish a fixed satellite attitude in inertial space. Spin stabilization is a passive method of fixing the attitude since, at least in the short term, no particular control action is needed to maintain a given attitude. There are other passive stabilization methods which take advantage of gravitational, geomagnetic or solar radiation force fields in space, but these forces are generally too weak for primary spacecraft stabilization at geostationary altitude.

Spin rates vary between 5 and 120 r.p.m. Low spin rates reduce the gyroscopic stiffness. Slow-spinning satellites are therefore more vulnerable to attitude drift under the influence of perturbing torques, although less propellant is needed to restore the attitude. High spin rates not only increase the stiffness, but also generate centrifugal forces which can cause structural and propellant flow problems. Spin rates in the range of 60–90 r.p.m. provide a happy medium.

Although spin stabilization is a particularly simple solution to the problem of fixing a satellite's attitude in inertial space, it has some disadvantages in terms of solar array utilization efficiency (Chapter 8).

Fig. 10.9. Definition of spacecraft rotation angles.

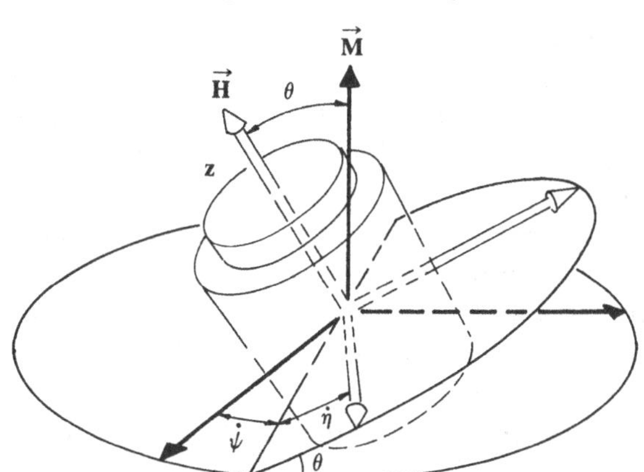

Moreover, high-gain communication antennae must be de-spun to stay pointed towards the earth (Chapter 12). Spin-stabilized spacecraft with de-spun structural members are referred to as *dual-spin satellites* (Fig. 10.2).

Active Stabilization

For geostationary spacecraft, the main alternative to passive spin stabilization is active *three-axis stabilization*. Satellites employing this method of stabilization remain motionless with respect to the satellite-earth range vector, apart from the solar array wings which complete one turn every day as they track the sun. Seen from inertial space, it is actually the spacecraft body which completes a daily turn to track the earth, while the array wings take a whole year to rotate 360°. Either way, these rotations are too slow to evoke any measurable gyroscopic effect.

If a three-axis-stabilized satellite had no angular momentum whatsoever, it would in fact not be stabilized at all, but would begin to tumble under the influence of perturbing forces such as those caused by the solar pressure. Tumbling is avoided by installing either spinning *momentum wheels* (Fig. 10.3) or *reaction wheels* (Fig. 10.4) inside the spacecraft body. In order to understand how they work, we must first define a set of orthogonal, body-referenced coordinate axes.

There is a geometric similarity between a three-axis-stabilized satellite and an aircraft as it flies above the earth (Fig. 10.10). The concepts of *pitch*, *roll* and *yaw* employed in aircraft control theory are also used to describe the three degrees of rotational freedom of a spacecraft. The pitch axis is parallel with the earth's north–south axis, the roll axis is aligned with the satellite velocity vector, and the yaw axis points to the centre of the earth.

Three-axis-stabilized satellites are more susceptible to perturbing torques in pitch than in roll or yaw. This is because the satellite, with all its protruding antennae and nozzles, is geometrically asymmetrical around the pitch axis and is therefore prone to be unbalanced by solar radiation forces. The resulting pitch torques vary in the course of the day as different satellite walls face the sun. These cyclic torques may be readily compensated if a *momentum wheel* (or, more accurately, a *momentum bias wheel*) is installed with its spin axis parallel to the satellite's pitch axis (Fig. 10.3). The wheel is maintained by an electric motor at a nominal speed (several thousand r.p.m.) such that a suitable momentum is attained. Gyroscopic stiffness provides resistance against perturbing forces in roll and yaw, whereas the slightest variation in wheel speed translates into a reactive movement of the spacecraft around the pitch axis.

By placing two separate wheels at a cant angle φ in the pitch–yaw plane (Fig. 10.11), it is possible to compensate for attitude errors in both pitch, roll and yaw. Pitch control is maintained by varying the two wheel speeds together, while yaw control is achieved by differential speed control. Let $\vec{\mathbf{H}}$ represent angular momentum. From Fig. 10.13:

$$H_{\text{pitch}} = (H_{\text{a}} + H_{\text{b}}) \cos \varphi,$$

$$H_{\text{yaw}} = (H_{\text{a}} - H_{\text{b}}) \sin \varphi$$

where $H_{\text{a}} = I\omega_{\text{a}}$ and $H_{\text{b}} = I\omega_{\text{b}}$. Correction in roll is made possible by *quarter orbit coupling*, since an error in yaw translates into an identical error in roll one-quarter of an orbit later (Fig. 10.12).

The gyroscopic stiffness in roll and yaw which comes with the use of a momentum wheel is expedient for satellite missions with moderate earth-pointing accuracy requirements (typically 0.2°), but it makes for rather sluggish pointing when higher accuracies are needed. In these circumstances it is often better to eliminate the momentum wheels and use *reaction wheels* instead (Fig. 10.4). While the former operate with a momentum bias (i.e. with wheel speeds varying around a nominal, non-zero value), the latter spin in either direction around zero speed. It is customary to mount one reaction wheel along each of the pitch, roll and yaw axes. This arrangement allows a satellite to be pointed with utmost

Fig. 10.10. Definition of roll, pitch and yaw for a three-axis-stabilized spacecraft.

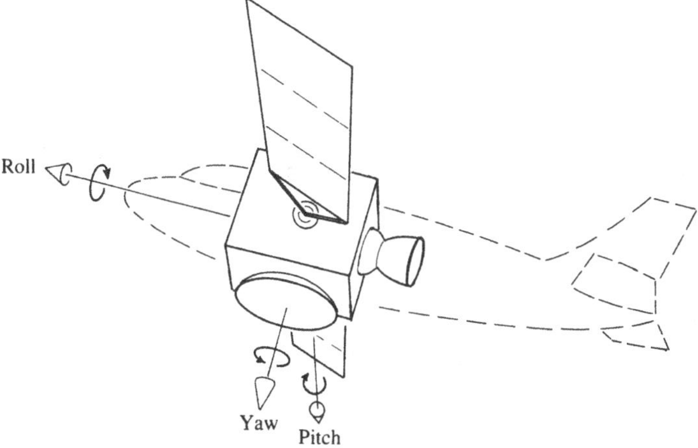

precision, though a price must be paid in terms of higher mass and greater complexity.

From a spacecraft dynamics point of view, the difference between a dual-spin satellite and a spacecraft equipped with a momentum wheel is more apparent than real. In the former case, part or all of the payload is de-spun while the platform is providing gyroscopic stiffness by virtue of its spin motion. In the latter case the entire spacecraft body is "despun" around the gyroscopically stiff momentum wheel. Even a three-axis-stabilized satellite can therefore experience nutation.

Attitude Correction

Spin and three-axis stabilization solve the problem of attitude fixation in the short term, but internal and external forces cause a satellite's attitude to wobble or drift, and these tendencies have to be corrected.

There are two kinds of force causing perturbing torques: *cyclic* (repetitive) and *secular* (non-repetitive). For example, a pivoting radiometer telescope (Chapter 13) constitutes an internal disturbance

Fig. 10.11. Pitch and roll/yaw control using canted momentum bias wheels.

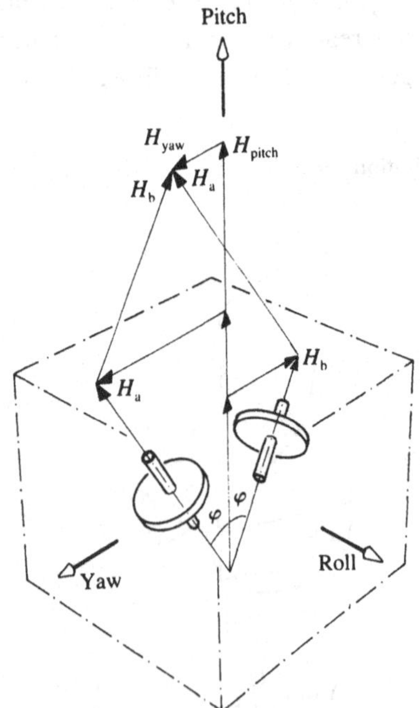

source and gives rise to cyclic torques. Friction in bearings between moving parts cause internal torques of a secular nature. Radiation pressure due to sunlight and albedo (sunlight reflected by the earth) creates external cyclic torques when illuminating asymmetric rotating spacecraft surfaces, while non-rotating asymmetries provoke external secular torques.

A *spin-stabilized* geostationary satellite is subjected to three different kinds of attitude disturbance: (i) nutation, (ii) instability and (iii) drift.

Nutation

We recall from the discussion under Gyroscopic Theory (p. 125) that nutation is caused by sudden, brief torques. Nutation may arise, for example, when a spinning satellite is released from the last stage of a launch vehicle by separation springs, or after a short thruster burn, or as a result of telescope pivoting. Provided the satellite is able to dissipate momentum energy (e.g. through structural flexing or internal fluid flow), it will gradually return to rotate around its major principal axis, i.e. the axis of greatest moment of inertia. Nutation can be eliminated more rapidly with the help of an energy-dissipating *passive nutation damper*. Most dampers consist of a slightly bent, sealed tube containing an inert

Fig. 10.12. Roll/yaw quarter-orbit coupling.

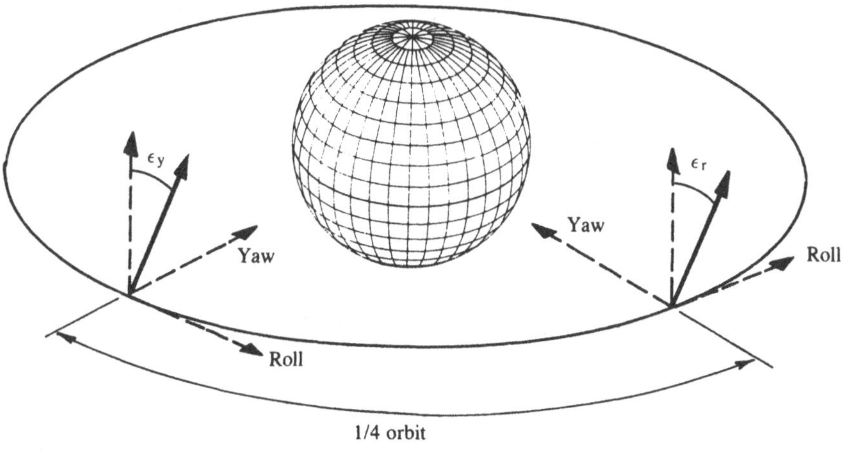

gas and a freely moving steel ball or liquid mercury. The tube is mounted parallel with the satellite spin axis. As the satellite nutates, the ball or the liquid experiences accelerations along the tube, whereby angular momentum is dissipated through the heat released on impact against the end stops, or through viscous friction with the gas.

Instability

Instability occurs when the satellite spin axis is different from the major principal axis. Left to its own devices, the satellite will nutate with increasing amplitude until the spin axis coincides with the major principal moment of inertia axis. If the major axis is one of the satellite's lateral geometric axes, we have a catastrophic situation known as *flat spin*. In the history of space flight, at least one satellite has been lost due to the miscalculation of the *moment of inertia ratio*. The ratio λ is defined as:

$$\lambda = 1 + \{(I_z/I_x - 1)(I_z/I_y - 1)\}^{1/2}$$

where I_z is the moment of inertia around the intended spin axis and I_x and I_y refer to the lateral principal axes. In the case of an axisymmetric spacecraft ($I_x = I_y$), $\lambda = I_z/I_x$. It is customary to design spinning satellites such that $\lambda > 1.1$ during all phases of the mission. There may be phases of relatively short duration, however, where this ratio cannot be readily achieved, e.g. before ejection of a perigee or apogee motor. In this case one has to resort to *active nutation damping*. Active damping is accomplished by pulse-firing an axial thruster at selected spin phase angles so as to "prop up" the satellite, thus preventing it from decaying into flat spin.

Attitude Drift

Attitude drift correction of a spin-stabilized satellite amounts to performing a small attitude manoeuvre. The techniques used will be discussed below under the section on Attitude control (p. 144).

Let us now examine how attitude disturbances on *three-axis-stabilized* satellites are corrected. We have already discussed how a satellite equipped with a momentum wheel absorbs cyclic perturbations in pitch. In addition to cyclic torques, there are secular pitch perturbations. They can also be removed by the momentum wheel, but this entails gradual spin-up or spin-down until the time comes when the wheel speed moves outside specified limits. Once the limit has been reached, i.e. when the wheel has *saturated*, it is necessary to bring the speed back to nominal by way of a *momentum dumping* operation. Dumping is performed by firing a thruster such that a pitch torque of suitable magnitude and orientation is applied to the satellite. The momentum wheel reacts instantly to this secular disturbance by changing its speed back to a value well within allowable limits.

A satellite which is controlled in pitch will eventually also drift in roll and yaw, and these drifts have to be corrected as well. Since a misalignment in yaw translates into an error in roll through quarter-orbit coupling, we need only be concerned about attitude correction in roll. The traditional correction method is to fire a thruster so as to "roll" the spacecraft back to nominal. A more elegant technique is to turn the solar wings slightly at an angle with each other such that the solar pressure creates a windmill effect.

Satellites equipped with reaction vessels are also subject to wheel saturation due to secular torques around all three axes. Momentum dumping of reaction wheels is carried out in much the same way as for momentum wheels.

Attitude Measurement

We have described earlier how satellites somewhat miraculously perform attitude correction, active nutation damping and momentum dumping. Whether the action is initiated by a human operator or by autonomous intelligence onboard a spacecraft, it is necessary to know the satellite's attitude so as to establish how it differs from a desired orientation. This knowledge is obtained through onboard attitude measurements using *attitude sensors*. Four different types of attitude sensors are commonly used on geostationary applications satellites: sun sensors, earth sensors, gyros and accelerometers.

Sun Sensors

Sun sensors consist of photocells which produce an electric current when illuminated by sunlight in the visible part of the spectrum (0.4–0.7 μm). The sun is an ideal target for attitude referencing due to its brightness and approximate point-source characteristics (seen from a geostationary satellite, the sun subtends 0.5°). In its simplest form, a sun sensor measures directly the output current I of a photocell and compares it with the maximum current I_{max} obtained when the solar incidence angle is perpendicular to the cell. The solar aspect offset angle θ_s' is then given by (Fig. 8.13):

$$I = I_{max} \cos \theta_s', \qquad \text{from which } \theta_s' = \cos^{-1}(I/I_{max}).$$

The angle θ_s' is highly ambiguous, however, as indicated in Fig. 10.13; it is also very inaccurate at sun angles near 90° due to the flatness of the cosine function. The ambiguity can be resolved and the accuracy improved if four photocells are grouped together and a mask is placed between the cells and the sunlight (Fig. 10.14).

The instrument just described constitutes an *analogue* sun sensor. It is well suited for three-axis-stabilized satellites. Other photocell arrangements and masking techniques exist in order to tailor the sun sensor field of view to the needs of particular three-axis-stabilized satellite missions.

Spin-stabilized satellites employ a different kind of analogue sun sensor, called the *V-slit sun sensor* (Fig. 10.15). It consists of photocells accommodated inside a housing through which a vertical and an inclined slit have been cut. The sun angle θ_s (= 90° − θ_s') is a direct

Fig. 10.13. Ambiguity of solar incidence angle measured by a single solar cell.

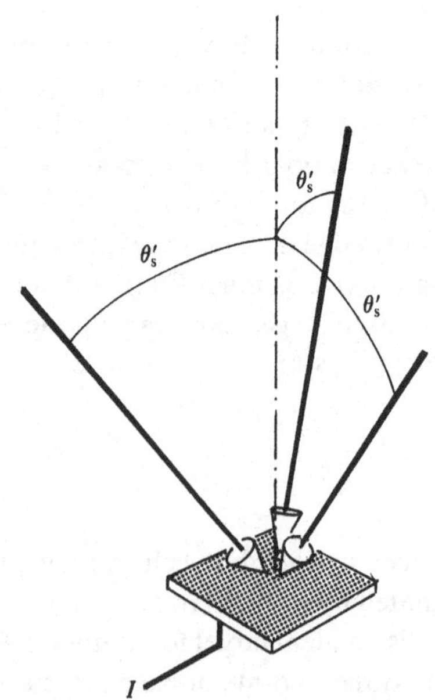

function of the satellite spin rate ω, the time Δt between sun sightings of the two slits, and the angle φ between the slits. From Fig. 10.16:

$$\tan \theta_s = \tan \varphi / \sin (\omega \, \Delta t).$$

Fig. 10.14. Ambiguity resolution of solar incidence angle using four solar cells and a mask.

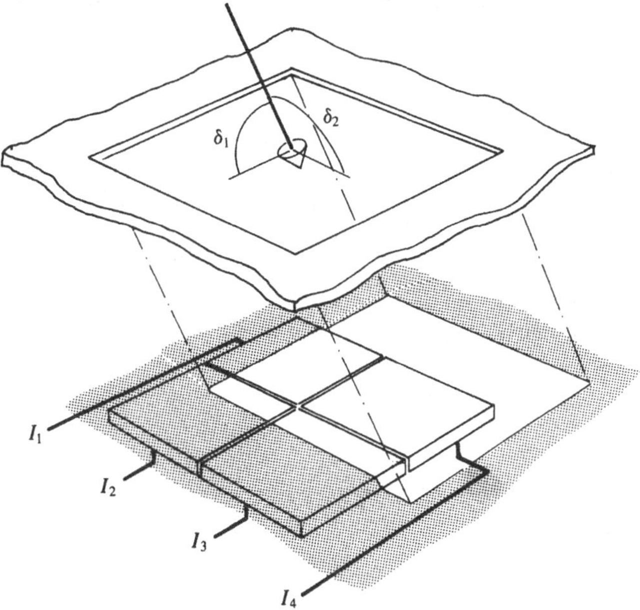

Fig. 10.15. V-slit analogue sun sensor.

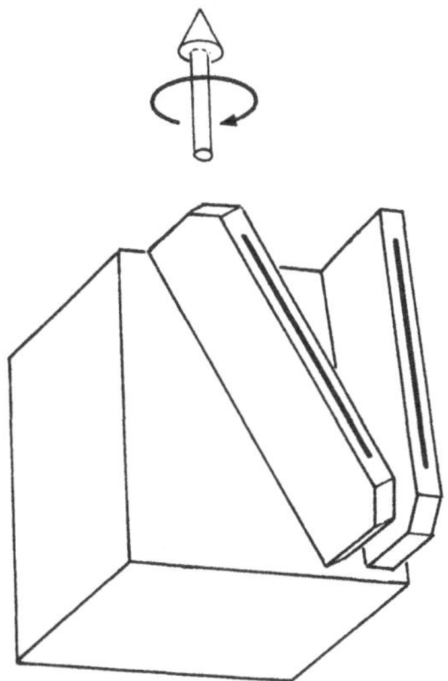

Some spin-stabilized satellites employ *digital* sun sensors of the type illustrated in Fig. 10.17. Sunlight passes through an entrance slit and subsequently via a refractive medium through a row of reticle slits before falling upon a bank of photocells. By laying out the reticle slits in a coded pattern on the sensor floor, the sunlight is made to stimulate different combinations of photocells at different solar aspect offset angles θ_s'. A digital read-out results which is suitable for direct telemetry transmission.

Earth Sensors

While sun sensors provide the angle between a satellite body reference vector and the satellite-sun vector, the satellite attitude

Fig. 10.16. V-slit sun sensor algorithm. The V-slit is projected onto a celestial sphere around the satellite's centre of gravity.

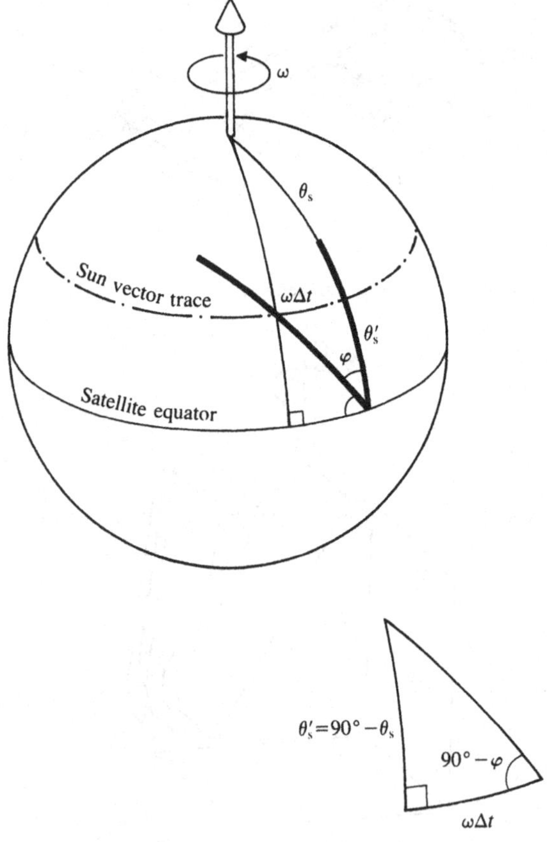

remains ambiguous since the body vector might be located anywhere along an ambiguity cone with a half-angle equal to θ_s. By establishing a similar ambiguity cone around the satellite-earth vector, the overall attitude ambiguity is reduced to the two intersections between the cones (Fig. 10.18). This residual ambiguity is resolved in the attitude determination process on the ground.

The earth-referenced ambiguity cone is provided by the earth sensor. Its half-angle is denoted θ_e. As a source of measurable radiation, the earth differs from the sun in two important aspects: it is much larger as seen from a geostationary satellite (17.4° instead of 0.5°), and its visible light goes out at night. The latter problem is solved allowing earth sensors to operate in the infrared (14–16 μm) spectral band rather than in the visible spectrum, given that the earth radiates heat regardless of solar illumination.

Finding the centre of the "large" earth is a little more complicated. Horizon scanners are used onboard spin-stabilized satellites. They scan the earth with two pencil-beam bolometers (Fig. 10.19) and record the transitions between space-earth and earth-space (Fig. 10.20). A bolometer is a very sensitive thermistor capable of measuring small thermal radiation changes, such as those occurring during scan transitions from space (4 K) to earth (250 K) and back to space. The earth angle θ_e is a function of the satellite spin rate ω, the time between horizon crossings

Fig. 10.17. Digital sun sensor, type Adcole.

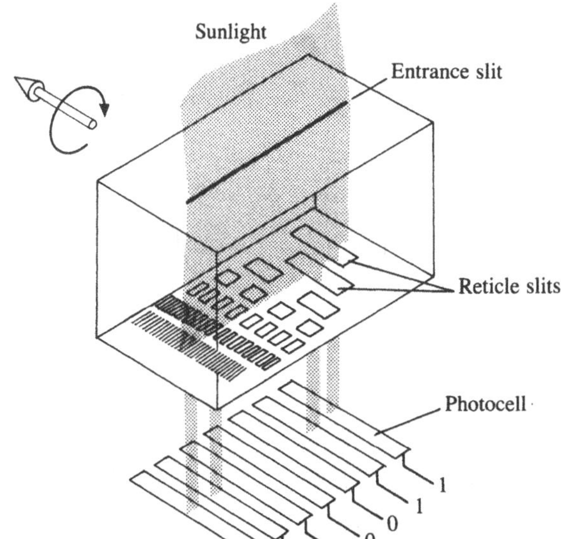

Δt_1 and Δt_2, and the angle 2α between the two bolometer pencil beams. From Fig. 10.21:

$$\cos\rho = \cos\theta_e \sin\alpha + \sin\theta_e \cos\alpha \cos(\tfrac{1}{2}\omega\Delta t_1),$$
$$\cos\rho = -\cos\theta_e \sin\alpha + \sin\theta_e \cos\alpha \cos(\tfrac{1}{2}\omega\Delta t_2)$$

and, therefore:

$$\tan\theta_e = \frac{2\tan\alpha}{\cos(\tfrac{1}{2}\omega\Delta t_2) - \cos(\tfrac{1}{2}\omega\Delta t_1)}. \tag{10.4}$$

Horizon scanners are also used on three-axis-stabilized satellites, but since these satellites do not spin, it is necessary to install earth scanning mirrors in the field of view of the sensors. Alternatively, *static earth sensors* may be used. These sensors eliminate the need for scanning mirrors and are therefore more reliable, albeit less accurate. They consist of thermopiles in a star configuration as shown in Fig. 10.22. The difference in radiance output between diametrically opposite thermopiles represents a measure of attitude offset from the satellite-earth vector, i.e. of the earth angle θ_e.

Having measured θ_s and θ_e, we are now in a position to determine the attitude of the satellite. In the case of a spin-stabilized satellite, the attitude is defined by the orientation of the spin axis **z**. Let **s** be the

Fig. 10.18. Attitude ambiguity cones of sun and earth sensor measurements.

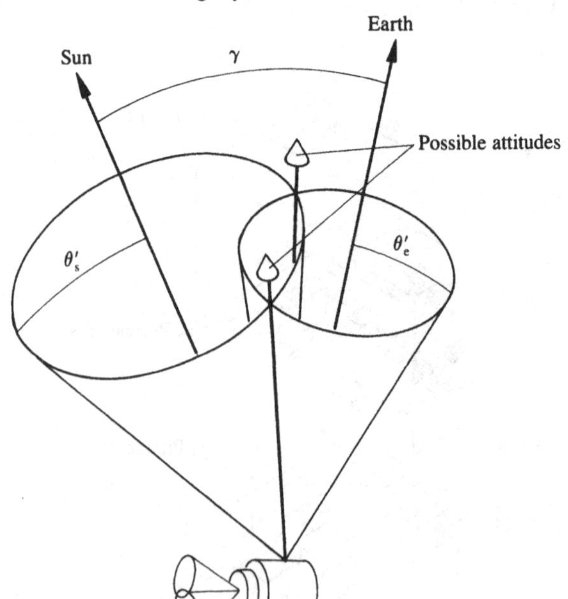

satellite/sun unit vector and **e** be the satellite/earth vector. If we select the coordinate system shown in Fig. 10.23, we have:

$$\mathbf{s} = (\sin\theta_s, 0, \cos\theta_s),$$

$$\mathbf{e} = (\sin\theta_e \cos\varphi, \sin\theta_e \sin\varphi, \cos\theta_e)$$

where φ is the angle between the planes containing **z–s** and **z–e**. φ is obtained by direct comparison of sun sensor and earth sensor read-outs.

Fig. 10.19. Pencil-beam bolometer earth sensor.

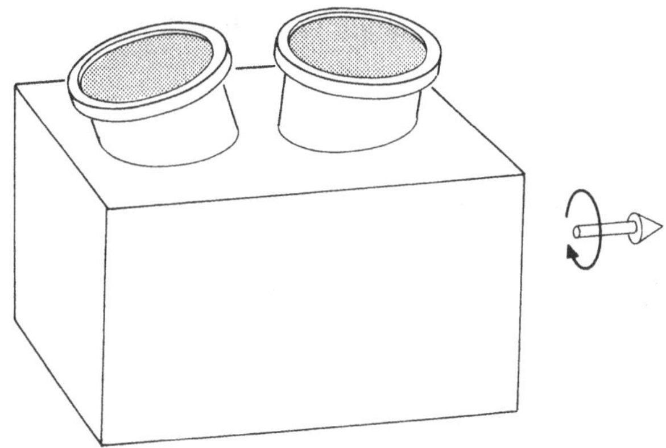

Fig. 10.20. Earth sensor horizon scanning.

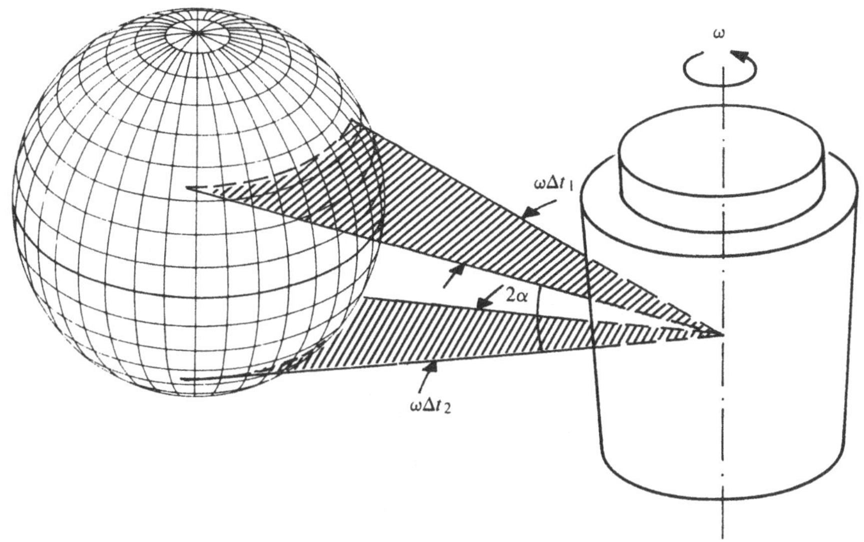

z is computed by solving the vector equation

$$\frac{\mathbf{z} \times \mathbf{s}}{\sin \theta_s} \times \frac{\mathbf{z} \times \mathbf{e}}{\sin \theta_e} = \mathbf{z} \sin \varphi$$

which yields:

$$\mathbf{z} = \mathbf{s} \cos \theta_s + (\mathbf{e} - \mathbf{s} \cos \gamma)(\cos \theta_e - \cos \theta_s \cos \gamma)/\sin^2\varphi \\ + (\mathbf{s} \times \mathbf{e}) \sin \theta_e \sin \theta_s / \sin \varphi \qquad (10.5)$$

where $\cos \gamma = \cos \theta_s \cos \theta_e + \sin \theta_s \sin \theta_e \cos \varphi$. Note that **s** and **e** may be defined in any suitable coordinate system, as long as they both belong to the same system. Equation 10.5 will then yield **z** in the chosen system.

Fig. 10.21. Horizon-scanning earth sensor algorithm.

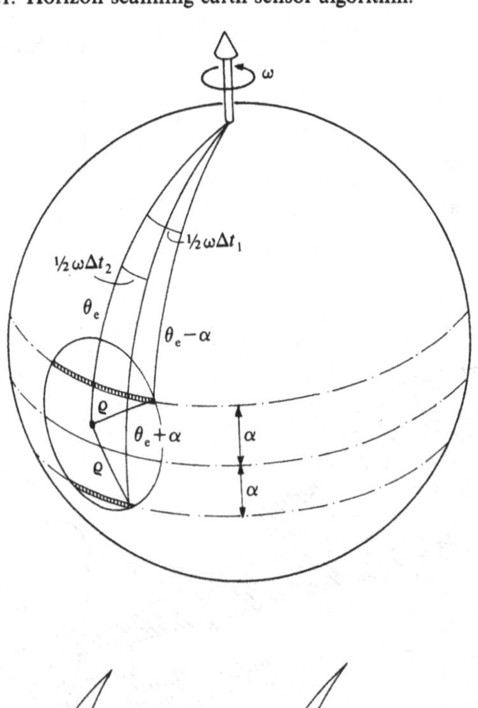

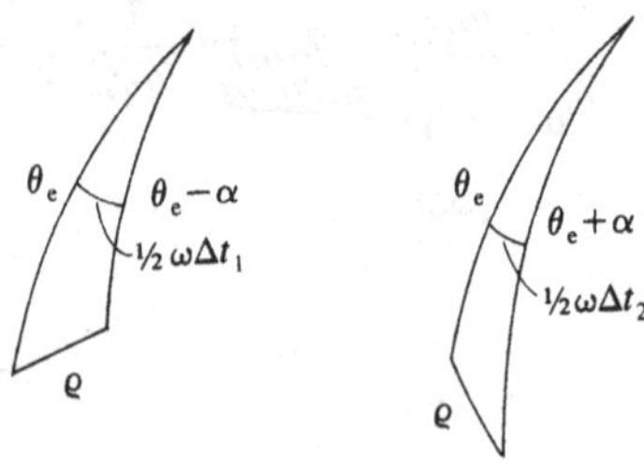

Gyros

We have seen earlier how three-axis-stabilized satellites correct cyclic and secular attitude drift by changing wheel speeds, or feathering solar wings in a windmill fashion. The timing of corrective action is determined by error signals from the sun and earth sensors. Since the drift is very slow, the correction control loops - whether on the ground or onboard the satellite - can afford to have long time constants.

During north–south or east–west station-keeping manoeuvres, however, the firing of thrusters causes rapid and violent attitude perturbations. Unless these are corrected almost immediately, there is a danger that the earth sensors might lose sight of the earth, leaving the satellite in an undefined attitude. It is therefore common practice to install small *gyros* along three orthogonal body axes, so as to obtain instant information on attitude perturbations in pitch, roll and yaw. The information is used by an onboard microprocessor to stabilize the spacecraft through modulation of the thrust.

Fig. 10.22. Static earth sensor for three-axis-stabilized spacecraft.

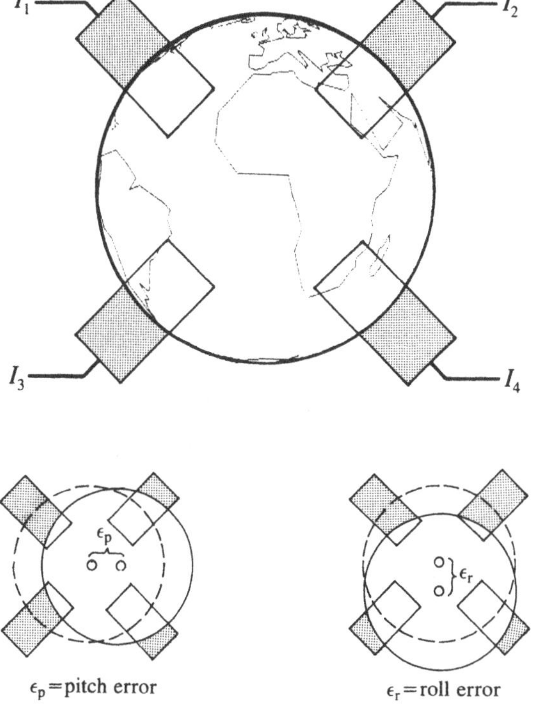

There are two kinds of gyros in common usage: rate gyros and rate integrating gyros. The functioning of a *rate gyro* (RG) is shown in Fig. 10.24. The perturbing torque $\vec{T}$ enters via the input axis. As the gimballed rotor tries to align its spin axis with the input axis, it is restrained by springs having a torque spring constant k. The springs limit the angular displacement of the spin axis to an angle θ which is proportional to the magnitude $\vec{T}$. The displacement angle θ is measured along the output axis.

Let $H = I_r\omega_r$ be the angular momentum of the rotor, and let ω_i be the angular velocity around the input axis resulting from $\vec{T}$. Then, in a steady-state environment:

$$k\theta = \omega_i H$$

or

$$\theta = \omega_i H/k.$$

Since θ is proportional to the angular *velocity* ω_i of attitude perturbations rather than to the angular *displacement*, a rate gyro cannot provide

Fig. 10.23. Coordinate system for attitude determination.

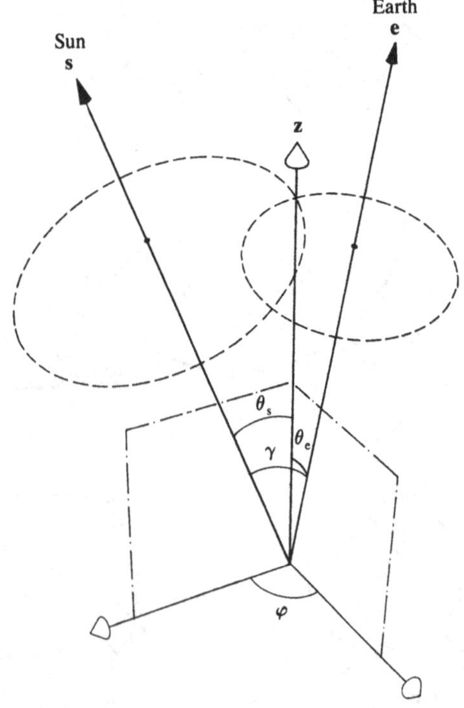

attitude error information directly. The RG output is, however, immediately usable for perturbation correction purposes.

A *rate integrating gyro* (RIG) differs from an RG in that the movement of the gimballed rotor is not restrained by springs. Consequently, the angular displacement of a RIG's output axis is a direct measure of any attitude drift component along the input axis. Attitude reconstitution is thereby simplified, but RIGs are more complex and expensive than RGs, and the gimbal motion is limited to only a few degrees.

Accelerometers

Although it is possible to detect and measure satellite nutation by examining the sinusoidal output of sun and earth sensors, a safer and more straightforward method is to employ *accelerometers*. An accelerometer is basically a mass or a pendulum restrained by springs. As the spacecraft nutates, the mass is subjected to variable centrifugal forces. The resulting amplitude and frequency of mass displacement is measured onboard the satellite and is used to modulate thruster duty cycles for active nutation damping.

Fig. 10.24. Principle of operation of an altitude gyro.

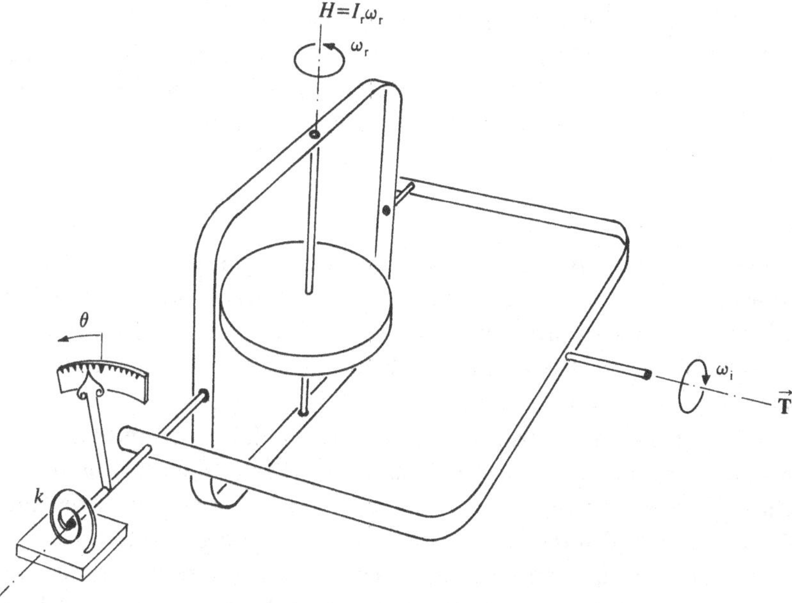

Attitude Control

So far we have covered attitude stabilization and measurement. We are now ready to study *attitude control*.

Geostationary applications satellites are designed to remain constantly pointed towards the earth. Apart from minor attitude corrections, the satellites do not require attitude reorientation as a rule. The main exception to the rule occurs when a satellite performs attitude manoeuvres during and immediately after the geostationary transfer orbit (GTO).

Spin-stabilized as well as three-axis-stabilized satellites normally spin in GTO, although a few spacecraft remain three-axis-stabilized throughout. Spin stabilization is a particularly simple method of maintaining satellite attitude during apogee engine firing and whenever ground station coverage is lost.

Attitude control of a spinning satellite is achieved by firing an axial thruster in pulsed mode. The pulses are phased with respect to some inertial reference – usually the output of a special sun sensor known as a sun presence detector – such that a well-defined torque vector is obtained. The spin-phase angle of the thrust determines the direction in which the satellite spin axis moves.

The tip of the spin axis vector draws a peculiar trace on the celestial sphere during an attitude manoeuvre. As the angle between the spin axis and the sun vector changes, the orientation of the torque vector in inertial space also changes. The resulting trace is neither a straight line nor a precession circle, but a logarithmic spiral or *rhumb line* with the satellite-sun vector in its centre (Fig. 10.25). A rhumb line forms a constant azimuth angle with the meridians of the celestial sphere. On earth, a ship or an airplane maintaining a constant bearing travels along a rhumb line.

The spin axis trajectory resembles the flight path of a moth approaching a lamp. Because the moth sees the light source by looking sideways, it flies towards the light along a spiral track. The moths's constant view angle is analogous to the constant phase-angle of the satellite thruster impulse.

Moving the spin axis from one attitude to another amounts to finding the correct rhumb line or *slew trajectory* on a celestial sphere (Fig. 10.26) whose pole is pointing towards the sun (if a sun sensor provides the phase reference). Manoeuvre planning is easier if the slew trajectory is plotted on a Mercator projection of the celestial sphere rather than on the sphere itself, because a rhumb line then becomes a straight line

(Fig. 10.27). Knowing the slope of the rhumb line, it is possible to calculate the required phase angle and number of thrust pulses. The calculated values are entered by telecommand from the ground into a data register onboard the satellite, whereupon the slew manoeuvre is executed automatically.

The curvilinear precession angle ϕ can be obtained if the thrust force $\vec{\mathbf{F}}$, the thrust moment arm $\vec{\mathbf{L}}$, the spin rate ω and the moment of inertia I are known. From Fig. 10.7:

$$\vec{\mathbf{T}} = \vec{\mathbf{L}} \times \vec{\mathbf{F}},$$

where $\vec{\mathbf{T}}$ is the control torque. The precession rate Ω of the spin axis is:

$$\Omega = \mathrm{d}\phi/\mathrm{d}t = \frac{L}{I\omega} F(t).$$

Fig. 10.25. Slew trajectory of a spin-stabilized satellite under the influence of an external torque.

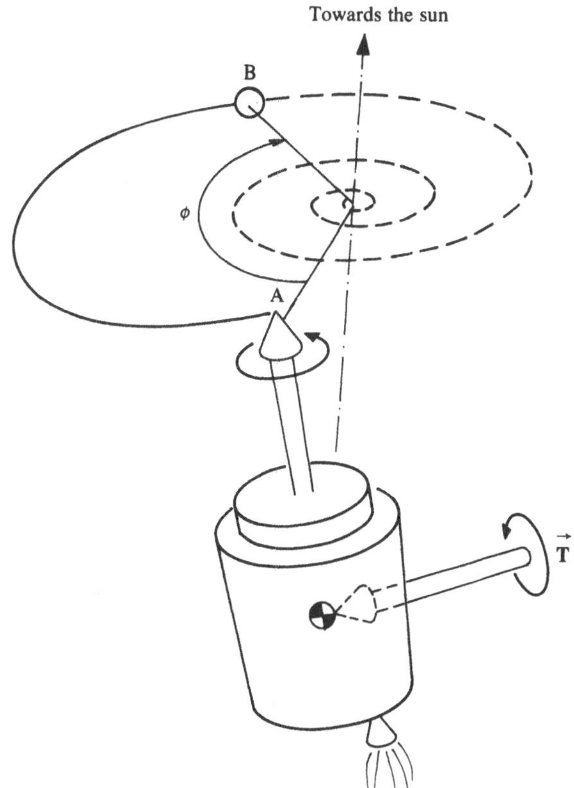

Then:

$$\phi = \frac{L}{I\omega}\int_0^t F(t)\,\mathrm{d}t \simeq \frac{L}{I\omega}\sum_{n=1}^{N} P(n)$$

where $P(n)$ is the control impulse number "n" due to $F(t)\,\mathrm{d}t$.

The attitude of a three-axis satellite which is spinning is GTO is

Fig. 10.26. Slew trajectory plotted as a rhumb line on a celestial sphere surrounding the spacecraft centre of gravity.

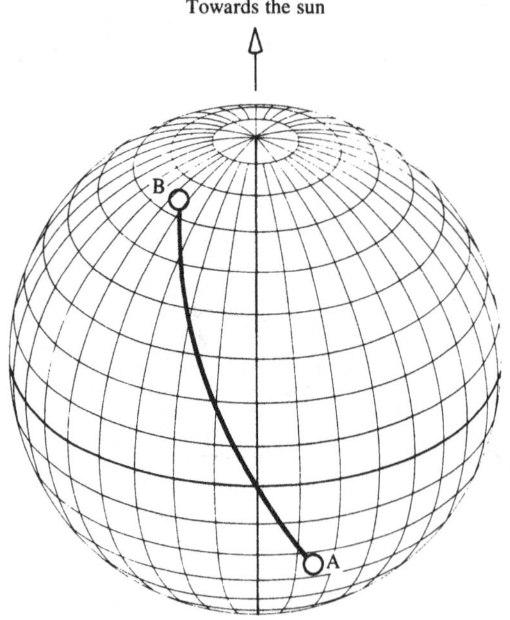

Fig. 10.27. The same rhumb line as in Fig. 10.26 plotted on a Mercator projection.

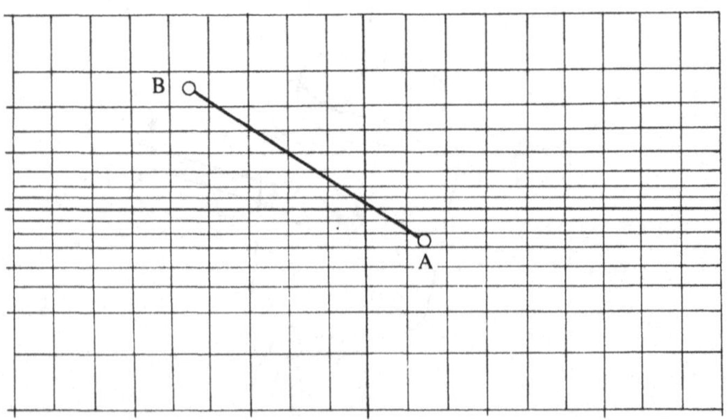

controlled in the same way as a spinner. Once the satellite has fired its apogee motor and is drifting slowly in a near-synchronous orbit, the time is ripe to despin it and point it towards the earth. Despinning is performed with the aid of tangential thrusters. Solar wing deployment follows. A wide-angle sun sensor helps the satellite to orient the roll axis towards the sun under thruster control such that power supply from the solar arrays is ensured. The satellite is then made to rotate slowly around the roll axis until the earth sensors along the yaw axis acquire the earth. Following earth acquisition, the momentum or reaction wheels are set in motion, and the satellite is ready for operational use.

Subsystem Architecture

Figure 10.28 shows the layout of the attitude control subsystem of a typical *spin-stabilized* satellite. A sun sensor and an earth sensor measure the elevation of the satellite-sun vector and the satellite-earth vector with respect to the satellite's equatorial plane. An accelerometer detects spacecraft nutation. The sensor outputs are converted to electric voltages and are processed in the Sensor Electronics Unit. The processed attitude signals are routed to the Antenna Despin Mechanism, the Thruster Control Electronics and the Telemetry and Telecommand Interface. The subsystem also includes a nutation damper.

The attitude control subsystem of a *three-axis-stabilized* satellite is more complex (Fig. 10.29). Assuming that the satellite is spin-stabilized

Fig. 10.28. Attitude control subsystem architecture for a spin-stabilized spacecraft.

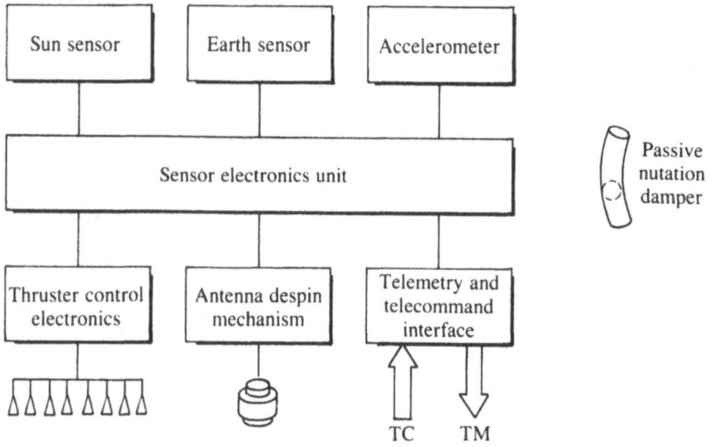

in transfer orbit, it requires one set of sun and earth sensors for the spinning phase, and another set for the three-axis-stabilized phase. In addition, the solar wings have their own sun sensors to facilitate sun pointing. A gyro and an accelerometer complete the sensor package.

The sensor outputs are converted and processed before being routed to the Telemetry and Telecommand Interface as well as to a microprocessor called the Control Law Electronics. Spacecraft attitude control instructions are issued to the Actuator Drive Unit which executes thruster pulses, turns the solar wings, and changes the spin speed of momentum or reaction wheels.

Bibliography

Hughes, P. C. (1986). *Spacecraft Attitude Dynamics*. New York: John Wiley.

Kaplan, M. H. (1976). *Modern Spacecraft Dynamics and Control*. New York: John Wiley.

Wertz, J. R. (1984). *Spacecraft Attitude Determination and Control*. Dordrecht: D. Reidel.

Fig. 10.29. Attitude control subsystem architecture for a three-axis-stabilized spacecraft.

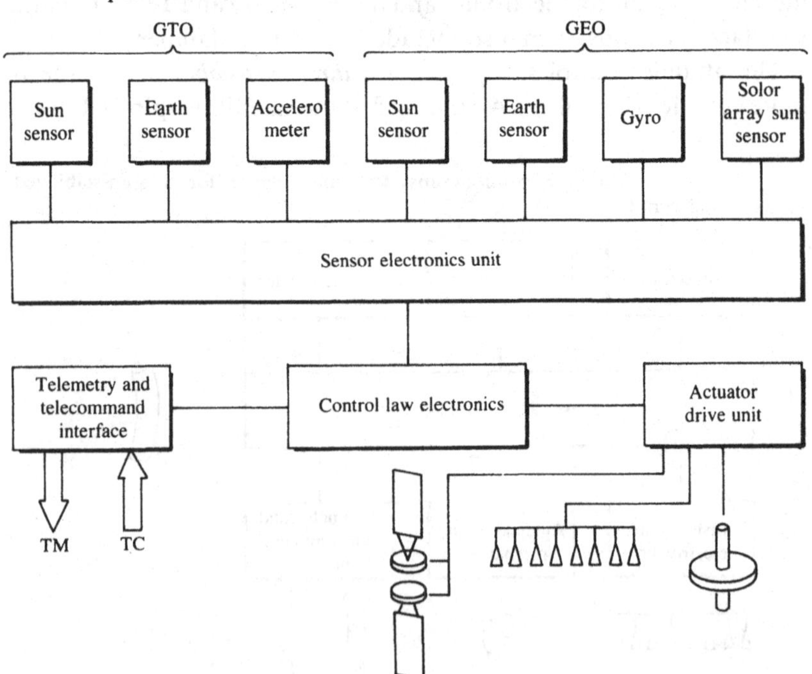

11

Telemetry, Tracking and Command (TT&C)

Introduction

Spacecraft controllers on the ground rely on telemetry to monitor the configuration and the health of a satellite. Telecommands provide the means to reconfigure, reorient and reposition the satellite by remote control. Tracking is a method of obtaining an update of the satellite's orbital elements.

The permanent loss of TT&C capability is the beginning of the end for a satellite. It will perhaps provide adequate service for hours, days or even weeks, but eventually it will drift away from its orbital location and nominal attitude. In the end it may degrade gracefully or suffer instant catastrophic failure as batteries discharge, propellants freeze, or electrical faults proliferate in the absence of corrective action from ground controllers.

Temporary loss of telemetry, telecommand and tracking is quite commonplace. Fortunately, the loss (called an "outage") is seldom the result of a satellite anomaly, but is usually triggered by a failure of the control centre computer, the ground station equipment or the data link connecting the two. Even so, spacecraft controllers are invariably seized by a feeling of impending doom when outages occur, and they act swiftly to restore contact before the satellite has come to harm.

Subsystem Architecture

Figure 11.1 shows the layout of a typical TT&C subsystem. Telecommands from the ground are picked up by a receiver. After demodulation they are passed on to a telecommand decoder. The decoder verifies the validity of commands, interprets their contents, and issues execution signals to the appropriate subsystem.

The status of satellite equipment is continuously monitored by sensors whose measurements are sampled sequentially by a telemetry encoder. These housekeeping measurements are converted and stored by the encoder before being released as a serial bit stream to the RF modulator and transmitter for onward transmission to the ground.

The telecommand receiver and the telemetry transmitter also serve as gateways for tracking signals in the form of ranging tones which originate on the ground and which consist of multiple sinusoidal waveforms. The signals are modulated onto a carrier by a ground station and are transmitted towards the satellite. Upon reception, the ranging transponder demodulates the tones, remodulates them onto a different carrier, and retransmits them to the originating ground station. The station compares the phase difference between the transmitted and received waveforms of each tone. From this comparison it is possible to establish the range (i.e. the distance) between the ground station and the satellite.

There is a growing trend to combine the traditionally hardwired telecommand decoder and telemetry encoder in a single software-driven microprocessor. Software offers greater flexibility to modify telecommand and telemetry functions, to diagnose internal circuit faults, and to take corrective action. Testing software is often more difficult than testing hardware, however, and making minor software changes late in the test programme can prove difficult.

Fig. 11.1. TT&C subsystem architecture.

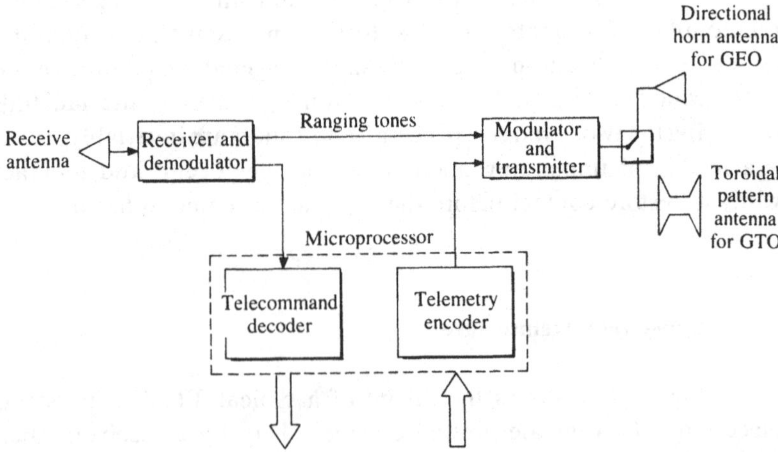

Telecommand

There are many ways of modulating and structuring a telecommand. Different space organizations have developed different standards. These norms are nevertheless all derived from a shared concern, namely that a command should be received only by the intended satellite and that, once received, it should be correctly executed. This concern is understandable because a corrupted command, or a spurious radio signal resembling a command, could have disastrous effects on the health of a satellite.

To minimize the risk of such mishaps, various space organizations have established telecommand standards which contain data redundancy and error detection features. One of the most widely used standards has been developed by the European Space Agency (ESA), and we will adopt this standard to illustrate the construction of telecommands. Before analysing the structure, it is useful to note that three different types of command are commonly used: (i) On/Off commands, (ii) 16-bit serial load commands ("memory load") and (iii) 24-bit serial load commands ("computer load").

On/Off commands are used for simple equipment reconfiguration, such as throwing a switch or resetting a register. *Memory load* commands occupy 16 bits and insert binary numbers into equipment registers and counters, e.g. the number of thruster pulses to be executed automatically. While 24-bit *computer load* commands are functionally similar to memory load commands, they are primarily used to program onboard microprocessors.

The general structure of standard ESA telecommand frame is shown in Fig. 11.2. The 16-bit *Address & Synchronization Word* (ASW) is a pseudo-random with a so-called Hamming distance of at least 3 from any other assigned ASW. The ASW contains the "address" of the target satellite, and the Hamming distance ensures that the risk of confusion among satellites is minimized even if the ASW is corrupted by radio frequency interference.

Fig. 11.2. ESA standard telecommand frame. Source: ESA.

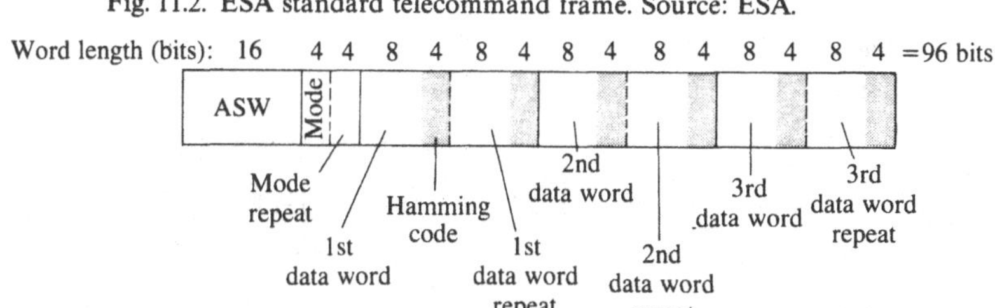

After the ASW follows a 4-bit *Mode Selection Word* which advises the decoder what type of command (On/Off, Memory Load, etc.) is being transmitted. The word is repeated once. If the two words are different (e.g. due to bit corruption), the decoder will reject the entire command, and the ground station has to transmit it once more.

The command instruction to be executed is contained in one or several of the three 12-bit *Data Words*. The actual instruction is made up of the first 8 bits, while the remaining 4 bits constitute an error-detecting (and sometimes error-correcting) Hamming code derived from the preceding 8-bit pattern. Each data word is repeated once. Thus the decoder has the opportunity to detect, correct and double-check the validity of each data word, and to reject the whole command frame if it is not satisfied.

A complete ESA command frame is thus made up of 96 bits. Frames may be linked together into command strings, whereby each successive frame begins with an ASW. The last frame in the string also has an ASW at the end. A preamble is always inserted before the first ASW. It consists of pure carrier transmission during 500 ms to stabilize the satellite receiver, followed by a 16-bit timing acquisition word (0000 0000 0000 0001) for onboard bit synchronization. A complete string of telecommands is shown in Fig. 11.3.

An On/Off command is made up of 8 bits, allowing up to three different commands to be included in a single frame (Fig. 11.4). A memory load command is accommodated by placing an equipment routing label in the first data word and spreading the 16-bit numerical content over the remaining two 8-bit words (Fig. 11.5). Computer load commands need all three available data words to hold 24-bit messages, leaving no room for routing labels (Fig. 11.6). Consequently, there can only be a single 24-bit computer load register onboard a satellite employing the ESA command standard.

Yet another type of telecommand, called a *Time-Tagged Command*, is used to link the execution of a command to a specific time determined by the satellite clock. Strictly speaking, time-tagged commands are not a

Fig. 11.3. String of telecommand frames. Source: ESA.

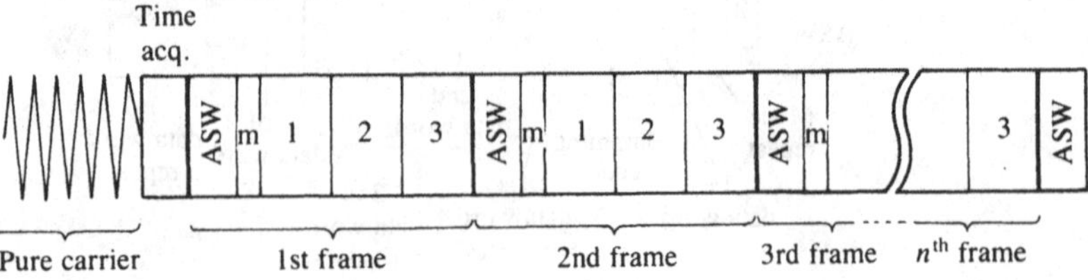

"type" by themselves, but they consist of a linked pair of frames in which the first frame is a memory load containing the time-tag, while the second frame is the actual command to be executed at the chosen moment. A special mode word number in the first frame is assigned to time-tagged commands, such that the decoder knows to associate it with the second frame. Time-tagged commands are used whenever a spacecraft function must be executed out of view of a ground station, e.g. to switch on mission-critical equipment or to fire a thruster in geostationary transfer orbit.

The care with which telecommands are protected through digital formatting may seem exaggerated at first glance, given that similar protection ought to be achievable by frequency and polarization discrimination, as well as by adopting different modulation techniques. In practice, however, these alternative protection measures are not always adequate or economical. Congestion in the radio spectrum has forced severe limitations in the width of frequency bands available for TT&C purposes. Some satellite owners are therefore encouraged to conserve available spectrum by operating their spacecraft on a common telecommand frequency and to rely on spatial separation among satellites as a means of reducing radio interference. Polarization discrimination, i.e. commanding a satellite in one polarization sense and another satellite in the opposite sense, offers very limited protection due to polarization and distorsion resulting from atmospheric rotation

Fig. 11.4. On/off telecommand. Source: ESA.

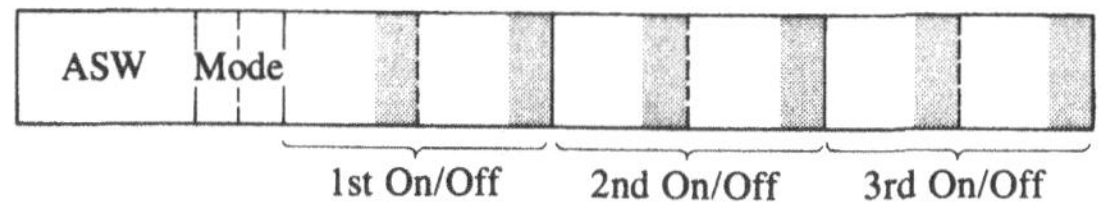

Fig. 11.5. Memory load telecommand. Source: ESA.

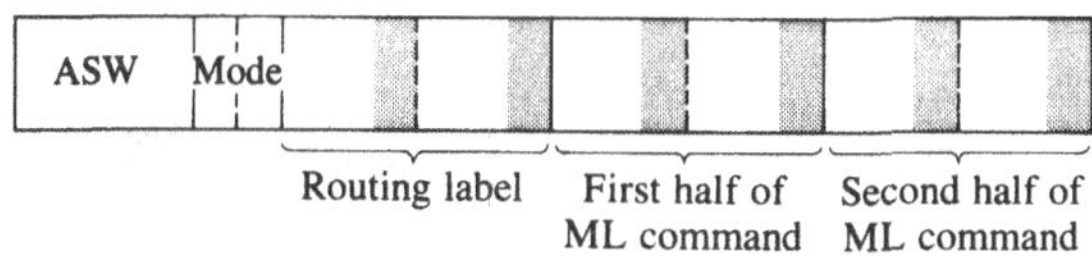

Fig. 11.6. Computer load telecommand. Source: ESA.

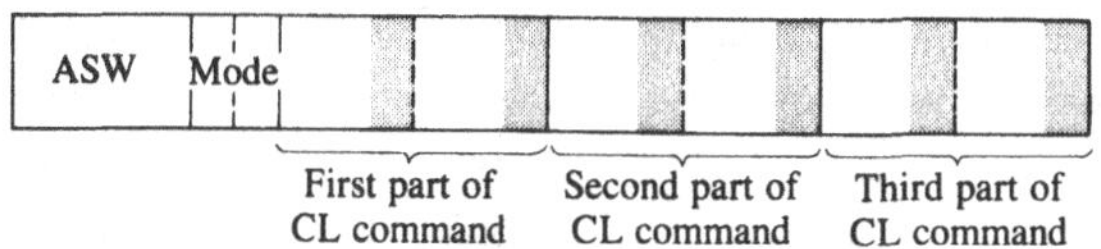

and from changes in spacecraft attitude. Adopting different modulation techniques, finally, defeats any attempt to optimize and standardize satellite as well as ground station telecommand equipment.

ESA telecommands are uplinked (i.e. transmitted from the ground to the satellite) in pulse code modulation (PCM) on an FM-modulated subcarrier which in turn FM- or PM-modulates the carrier. Telecommand receivers of most geostationary applications satellites are tuned to S-band (2 GHz) or C-band (6 GHz). The transmission bit rate varies between 90 and 1500 b.p.s.

Telemetry

Broadly speaking, receiving telemetry from a satellite is the reverse process of sending telecommands. Telemetry data is downlinked from the satellite and arrives at the ground station in the form of a PCM-modulated serial bit stream. The telemetry structure is quite different, however, from that of a telecommand.

A block of telemetry data is known in ESA parlance as a *format* (Fig. 11.7). The format is subdivided into *frames*, and each frame is made up of 8-bit *channels* or *words* which contain information about the satellite's status and health. Satellite designers are free, within certain limits, to choose the number of frames per format and the number of words per frame, and also the bit rate. A typical geostationary appli-

Fig. 11.7. Typical telemetry format.

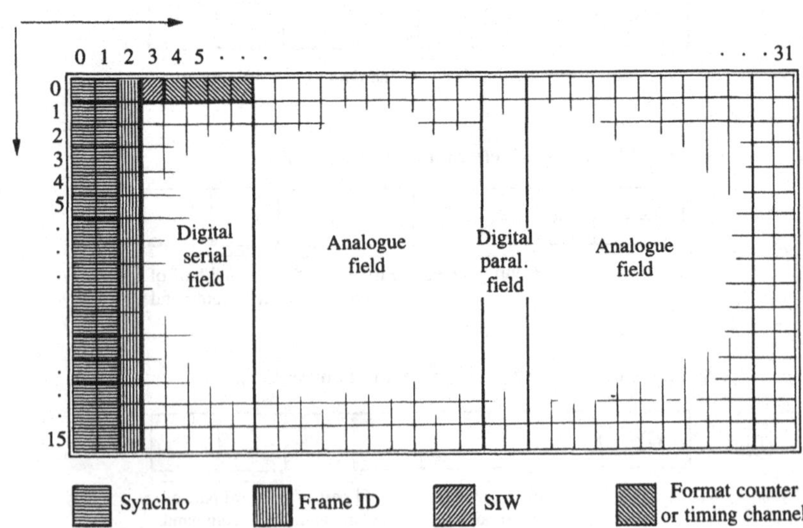

cations satellite might transmit a format composed of 16 frames by 32 words to the ground at a rate of 410 b.p.s. In this example a complete format would be transmitted every $16 \times 32 \times 8/410 = 10$ s.

Somewhere near the beginning of each updated format, a *Synchronization & Identification Word* (SIW) containing a fixed code identifies the satellite originating the telemetry. The SIW is the telemetry equivalent of the telecommand ASW.

Each frame is preceded by a 16-, 24- or 32-bit pseudo-random synchronization code and an 8-bit frame identification word which allow the ground station reception equipment to find its way through the format. Keeping a record of formats is facilitated by the provision of a *format counter* comprised of between one and four adjacent 8-bit words in the first frame. The value of the counter changes incrementally from one format to the next; alternatively it serves as a *timing channel* by providing a binary read-out of the onboard clock.

The telemetry channels described so far represent overhead functions needed by the ground station equipment to synchronize on the telemetry, and by the control centre computers to decommutate individual words. The remaining channels, or words, constitute the telemetry proper. The words themselves are digital representations of analogue or digital measurement samples. Analogue parameters such as spacecraft voltages, currents, power, radiance levels, pressures and temperatures are converted by transducers to voltages, usually in the range 0–5 V. These analogue voltages are subsequently translated by the telemetry encoder to digital values in the range 0–255, corresponding to the 8 bits in a word. Such a word is called an *analogue* channel, where "analogue" refers to the origin of the measurement rather than to the word itself which of course is digital. *Serial digital* channels comprise one or several adjacent 8-bit words which provide direct read-out of binary registers, counters and clocks. *Parallel digital* channels are functionally similar to the serial variety, but the binary information is shifted in parallel rather than serially between data registers. *Digital bi-level* channels contain 8- or 16-bit clusters of On/Off status bits, where each bit indicates how a particular equipment is switched in the overall electrical circuit. *Datation* channels are 8- or 16-bit digital channels showing the time, according to the spacecraft clock, when a particular automatic event took place onboard the spacecraft (e.g. at what time a sun pulse was delivered by a sun sensor).

The spacecraft clock consists of a crystal oscillator driving a simple binary counter. The content of the counter is sampled in the timing channel. By correlating the content of the timing channel with its precise arrival time at the ground station receiver, it is possible for the station computer to link spacecraft time with Greenwich Mean Time. This linkage is necessary for the archiving and retrieval of telemetry formats, and also for manual telecommanding of time-critical manoeuvres. For example, knowing the GMT of pulse delivery from a sun sensor could be crucial for manual attitude manoeuvring of a spin-stabilized satellite.

In the example we gave earlier, a 16-frame, 32-word format was transmitted every 10 s. If a particular measurement is only sampled by the encoder once per format, then the sampling rate is once per 10 s. This might not be frequent enough for some measurements where higher time resolution is sought. The spacecraft designer then has the option to *frame commutate* a particular word, i.e. the word is repeated once per frame instead of once per format. Alternatively, he may allow a word to be sampled less often than once per frame but more than once per format, in which case the word is said to be *subcommutated.* The designer may also *supercommutate* a telemetry word by allowing it to appear more than once per frame.

In the ESA standard, the telemetry data stream is pulse-code modulated. The data may be either loaded directly on a transmit carrier or modulate a square-wave subcarrier first. To minimize interference between telemetry transmission and telecommand reception, the two signals are assigned opposite polarizations.

Tracking

A satellite is regularly "tracked", i.e. the distance to the spacecraft is measured in order to derive an update of the orbital elements. By measuring variations of the range vector $\vec{\mathbf{r}}$ over time, and with the aid of Eqns 2.1–2.10, it is possible to establish a set of equations from which the six standard elements a, e, i, Ω, ω, ν can be derived.

Tracking and orbit determination are ground-based activities, and as such they are outside the scope of the present book. Since satellites assist in relaying tracking signals, we will nevertheless explore the most common method used in determining the distance to a geostationary applications satellite. The method is called *tone ranging.* A ground station sequentially modulates the telemetry carrier with low-frequency sinusoidal tones. Each tone is received, demodulated, remodulated and retransmitted on the telecommand carrier to the originating ground station by the satellite. By measuring the phase difference ϕ between the original and the returned tones, the station is able to determine the propagation delay and hence the two-way travel distance.

The higher the tone, i.e. the shorter the wavelength λ, the more accurate the range determination d, since:

$$d = \tfrac{1}{2}(\phi/360° + 2n)\,\lambda$$

where n is an ambiguity factor (Fig. 11.8). For example, the wavelength of a 28-KHz tone is 10 700 m. Such a tone would allow the satellite position to be determined with an accuracy better than 10 700/2 = 5400 m, disregarding the ambiguity factor n for the moment. If the phase delay measurement equipment at the ground station is capable of resolving phase differences of 1°, then the theoretical accuracy of the distance measurement becomes 5400/360 = 15 m. In practice, the accuracy is degraded by RF noise and by phase-distorting equipment imperfections both on the ground and onboard the satellite.

The ambiguity n is resolved by transmitting a succession of lower tones. Using the same calculation method as above, we find that a 4-Hz

Fig. 11.8. Orbit determination by tone ranging.

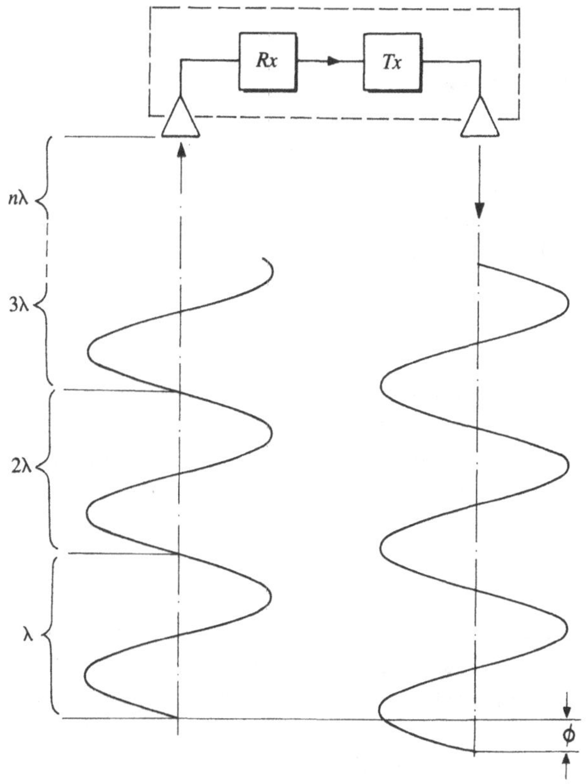

tone has a wavelength of 75 000 km, so that a geostationary satellite's distance can be determined within 75 000/2 = 37 500 km without any ambiguity whatsoever, although the accuracy of 100 km is totally unacceptable. Three or four intermediate tones, coupled with prior knowledge of the satellite's approximate whereabouts, suffice to improve the accuracy and resolve remaining ambiguities.

The ESA standard for ranging at S-band (2 GHz) is actually made up of a family of seven tones, namely, 8, 32, 160, 800, 4000, 20 000 and 100 000 Hz. These tones are designed for arbitrary earth orbits, not just geostationary. Apart from resolving ambiguities, the intermediate tones are employed to determine elliptic orbits, such as the geostationary transfer orbit, where the range distance to the satellite undergoes rapid changes.

The solution of orbit determination equations is facilitated by supplementing range measurements with angular information obtained from the TT&C antenna steering mechanism at the ground station. The antenna is equipped with an autotrack facility which allows it to home in automatically on the telemetry carrier transmitted by the spacecraft.

TT&C Antennae

The design of TT&C antennae for geostationary satellites is largely dictated by transfer orbit operations. To ascertain that the satellite has survived the rough ride into space, and to prepare attitude as well as orbital manoeuvres, spacecraft controllers need rapid access to telemetry and ranging data. TT&C operations are virtually excluded near the perigee where satellite/ground contact tends to be very brief due to the high orbital speed and low altitude of the spacecraft. Consequently, most manoeuvres must be prepared and executed closer to the apogee. This coverage limitation imposes a severe time pressure on the spacecraft controllers, and it is therefore essential that they have continuous TT&C access to the satellite for as long as it is in view of the ground stations.

Because the satellite is rotating while moving across the sky and also keeps changing its attitude, the station-to-satellite vector moves through all the quadrants of the satellite's geometrical reference frame. Consequently, a prerequisite for continuous TT&C access is that the satellite's transmit and receive antennae are *omnidirectional*, i.e. that their radiation pattern describes an approximate sphere around the spacecraft. Perfectly spherical patterns are difficult to achieve due to interference caused by the proximity of the spacecraft body, but an adequate pattern may be obtained by arranging quarter-wavelength monopole antennae in a *turnstile* configuration. Alternatively, it is possible to construct a quasi-omnidirectional pattern with a *toroidal* or a *cardioidal* antenna supplemented by conical *horn* antennae, each of which is *directional* (i.e. radiates in a preferential direction). Some of these antenna types are illustrated along with communication antennae in Chapter 12.

Most of the RF power radiated by a telemetry transmitter through an omnidirectional antenna is wasted, since only a small fraction travels along the line-of-sight to the ground station. With the transmitter efficiency limited to 40–70%, a waste of RF power translates into an even greater waste of electric power generated onboard the satellite. Such squandering of precious electricity may be acceptable in transfer orbit when the payload is idle, but once the satellite reaches geostationary orbit and the payloads are switched on, omnidirectional telemetry transmission can no longer be afforded. The solution is to switch the telemetry to a single directional antenna aimed at the earth, thereby enabling the antenna *gain* to make up for a reduction in transmitter output power. This directional antenna could be a *parabolic dish* or an *element array* which is shared with a communications payload through frequency multiplexing; alternatively, a dedicated horn antenna may be used.

A more detailed description of different antenna types is found in Chapter 12.

Bibliography

PCM Telecommand Standard (1978). ESA PSS-45 (TTC.A.01), Issue 1. Noordwijk: European Space Research and Technology Centre.

PCM Telemetry Standard (1978). ESA PSS-46 (TTC.A.02), Issue 1. Noordwijk: European Space Research and Technology Centre.

Ranging Standard (1980). TTC-A-04, Issue 1. Noordwijk: European Space Research and Technology Centre.

12

Communications Payload

Introduction

In Chapter 5–11 we covered the classical spacecraft *platform* subsystems. The present chapter introduces the *payload* section of this book. The payloads of interest in geostationary satellite applications perform the functions of communication (this Chapter) and meteorological imagery (Chapter 13).

All geostationary applications satellites communicate information, be it language, pictures or abstract numbers. When we talk about communications satellites, however, we limit the definition to spacecraft which *relay* voice, data, telex, facsimile, or television pictures from one part of the earth to another, as opposed to telemetry and imagery which *originate* in the satellites themselves.

Satellite communications is a new, vast and fascinating engineering science. Along with orbital mechanics and attitude and orbit control, it is also one of the most "difficult" spacecraft subjects to understand, mainly because the student is faced with a wealth of mathematical abstractions from the outset. In the present chapter we will provide an outline of basic communications theory and how it influences the architecture of a communications payload. The interested reader is advised to consult the books listed in the Bibliography for a more in-depth coverage of the subject. The list is far from exhaustive, and suggestions for further reading are made in the Reference sections of those books.

Transmission Capacity Versus Power and Bandwidth

We will begin this chapter by examining how various transmission techniques allow a designer to optimize the communications capacity of a satellite by trading off two precious commodities, namely spacecraft power and frequency bandwidth. Spacecraft power is in short supply because of the way in which it translates into mass, and hence launch cost. Frequency bandwidth has to be limited to prevent signals from being drowned by background noise, and to avoid interference among the growing number of spectrum users.

Any RF transmission consists of three components: carrier, signal and noise. The quality of a satellite communications link is, in the first instance, determined by the *carrier-to-noise ratio* (*C/N*) at the receiver on the ground or onboard the satellite. In the case of analogue modulation (i.e. amplitude, frequency or phase modulation), the post-detection *signal-to-noise ratio (S/N)* is the ultimate quality indicator while, in digital modulation, quality is measured in terms of bit error rate (*BER*). Both *S/N* and *BER* are functions of *C/N*, as illustrated in Figs 12.1 and 12.2.

The carrier (*C*) is a pure radioelectric sinusoid which carries no information by itself but which, when modulated by a signal, breaks up into a number of *sidebands* containing the signal intelligence (Fig. 12.3).

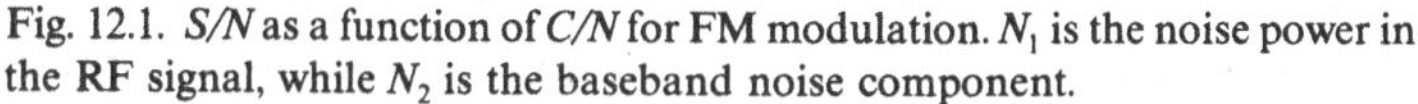

Fig. 12.1. *S/N* as a function of *C/N* for FM modulation. N_1 is the noise power in the RF signal, while N_2 is the baseband noise component.

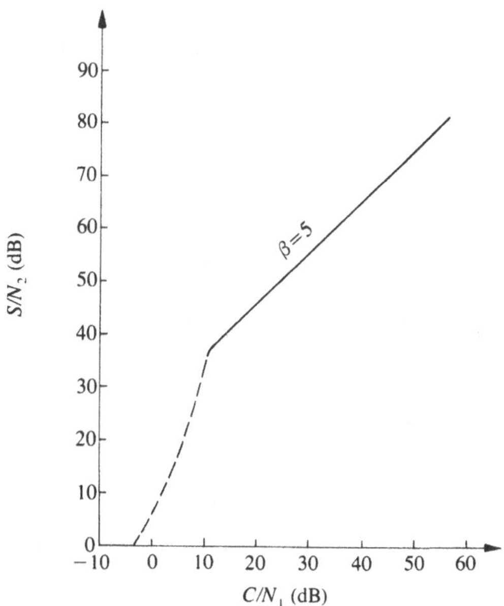

The carrier power level C at the receiver input is sometimes denoted P_r and measures a few pico-Watts (pW); it is all that remains of the transmitted power P_t (usually several hundred or thousand Watts) after the 37 000–42 000-km journey through space. By signal (S) we understand the part of the transmission which carries intelligible information. It starts out as an analogue (e.g. speech) or digital (e.g. data) information stream occupying a finite part of the frequency spectrum (baseband). The signal is made to "ride" on the carrier through modulation and is recovered after demodulation (Fig. 12.4). Noise (N), finally, is an undesirable and more or less random power ingredient which corrupts the intelligibility of the signal. For example, thermal noise is generated by resistive devices in electronic circuits and by background radiation picked up by the main and side lobes of antennae; the amount of thermal noise power is directly proportional to the temperature of the device or the environment according to:

$$N = kTB \tag{12.1}$$

where k is Boltzmann's constant = 1.38×10^{-23} J/K, T is the temperature in degrees Kelvin and B is the bandwidth in Hertz. Interferences due to intermodulation and extraneous signal traffic can often be converted to equivalent thermal noise. Thermal noise power levels from various sources may be added linearly to obtain the total noise figure.

The intelligibility of the signal S after demodulation depends on how strongly it rises above the total noise N. The ratio S/N is therefore an

Fig. 12.2. *BER* as a function of *C/N* for digital modulation.

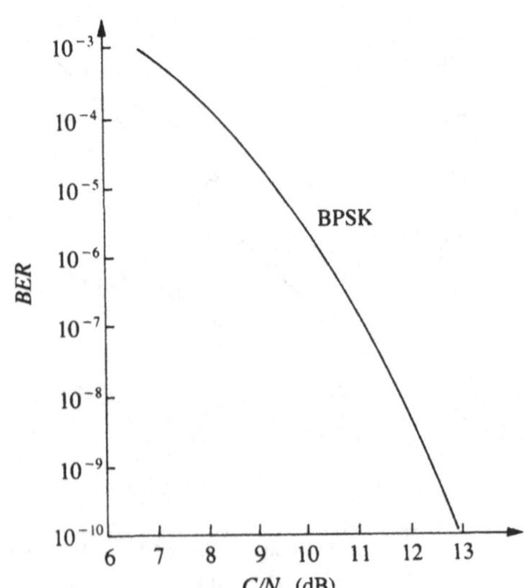

important independent parameter when designing space communications systems.

A satellite transponder can be likened to a "bent pipe" in space, through which a flow of information is to be transported. Because satellites are expensive to build and launch (typically \$100 m–\$400 m each), it is desirable to maximize revenue by using the transponder to the very limit of its throughput capacity. Just as the capacity of a water pipe is limited by its diameter and ability to withstand pressure, a transponder channel is limited by its frequency bandwidth and ability to digest a rapid flow of information. Let us examine these limitations in detail.

Take data relay as an example. Suppose there is a requirement to relay symbols M made up of m measurement levels (e.g. telemetry words of the kind described in Chapter 11). The symbols are to be transmitted at a rate $\dot{M} = \mathrm{d}M/\mathrm{d}t$. In order to manage m levels, each symbol must

Fig. 12.3. AM and FM in the frequency domain. Typical sidebands and required transformation bandwidth resulting from carrier modulation by a pure sinusoid.

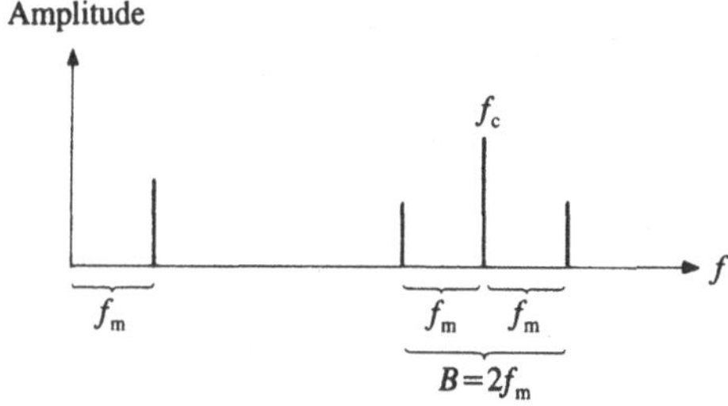

Amplitude modulation

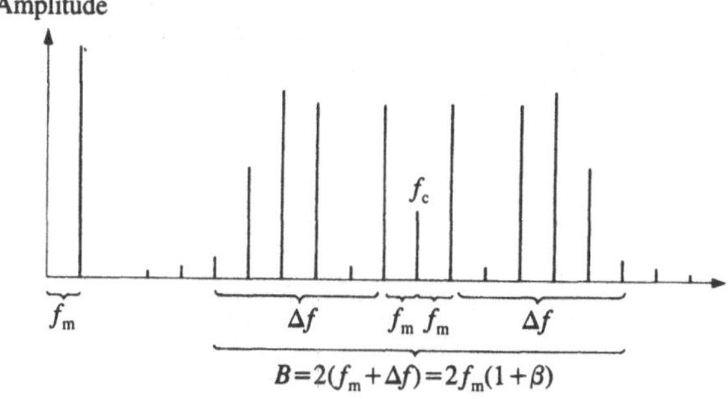

Frequency modulation

contain $D = \log_2 m$ binary data bits (in the case of housekeeping telemetry, $D = 8$ since $m = 256 = 2^8$). If the symbol rate is $\dot{M}$, it follows that the data rate, or *bit rate* $\dot{D}$ is equal to:

$$\dot{D} = \dot{M} \log_2 m. \tag{12.2}$$

According to Shannon (1948), the maximum information capacity Q (in our case the maximum bit rate $\dot{D}$) that can be relayed through a transponder channel is a function of two parameters, namely the channel frequency bandwidth B and the specified signal-to-noise ratio S/N. Thus:

$$Q = \dot{D} \leqslant B \log_2(1 + S/N). \tag{12.3}$$

The *Nyquist theorem* states that:

$$\dot{M} \leqslant 2B. \tag{12.4}$$

By combining Eqns 12.2, 12.3 and 12.4, we find that:

$$m \leqslant (1 + S/N)^{1/2}. \tag{12.5}$$

As we shall see later, it is convenient to re-format Eqns 12.4 and 12.5 as follows:

$$\dot{M}/2 \leqslant B, \tag{12.6}$$

$$2 \log_2 m \leqslant \log_2(1 + S/N). \tag{12.7}$$

Fig. 12.4. AM and FM in the time domain with the carrier modulated by a pure sinusoid.

Baseband

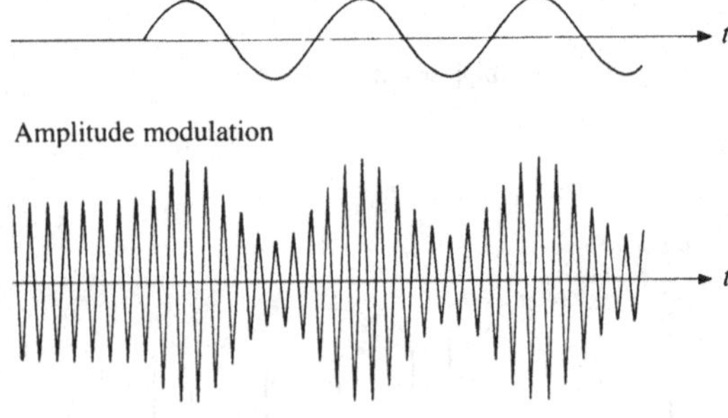

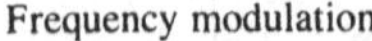

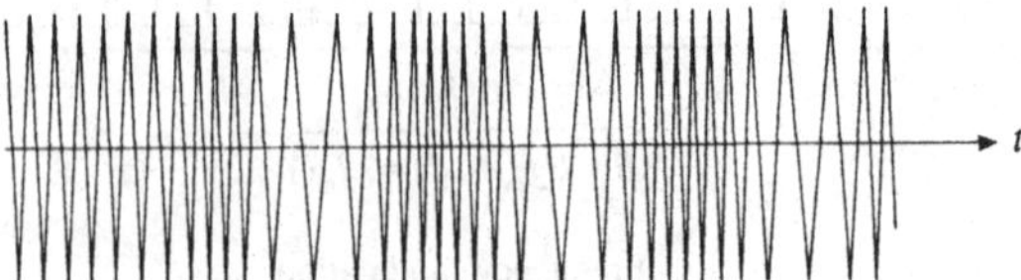

Our communication system requirement can now be restated as follows. A data package containing a total of D bits is to be relayed within a time period T via a "bent pipe" in the form of a satellite transponder channel having a maximum capacity Q. A specified signal quality S/N shall be achievable after demodulation.

Fontolliet (1986) has conceived an illustration of the above requirement (Fig. 12.5). The box-shaped package D has to fit inside the rectangular cross-section of the transponder channel. The shape of the data box may be altered by changing either the symbol rate $\dot{M}$, the number of measurement levels m, or the period T. Once the shape of the box has been decided, it remains to adapt the channel capacity Q by trading off the bandwidth B against the signal power S. This trade-off is precisely the transponder designer's main preoccupation which we discussed earlier.

The channel bandwidth must be wide enough to accommodate most of the energy contained in the modulation sidebands, but not be so wide that the signal is drowned by noise (Eqn 12.1) or by extraneous transmissions on adjacent frequencies. The S/N can be improved by a judicious choice of encoding and modulation techniques, and by adopting the latest in low-noise component technology.

Fig. 12.5. Digital data packaging to fit within the transmission capacity of a satellite transponder. Adapted by permission, from Pierre-Girard Fontolliet, *Telecommunication Systems*, p. 146. © 1986 Artech House Inc.

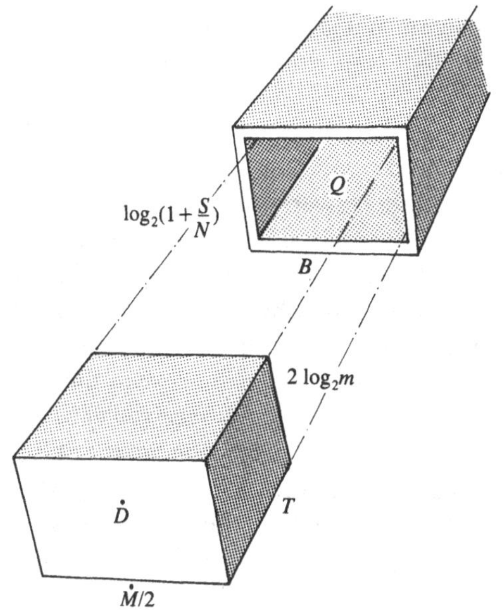

Equations 12.1–12.7 are theoretical only and paint a very simplistic picture of the real constraints influencing the design of space communications systems. Aberrations in the radio transparency of the atmosphere and in the transfer function of electronic circuits make these constraints more severe than the above equations would indicate. The manner in which satellite transponders are accessed from the ground also serves to complicate matters. For example, a common method of maximizing transponder value for money is to broaden the bandwidth quite considerably and to fill the increased capacity with signals interleaved in time or travelling on different carrier frequencies. The former method is called *time division multiple access* (TDMA), the latter *frequency division multiple access* (FDMA). These techniques, along with others aimed at enhancing communications capacity, render the mathematical analysis of system design far more complicated than in the foregoing example.

Subsystem Architecture

The primary elements of a satellite communications subsystem are the receiver, the transmitter and the associated antennae (Fig. 12.6). The transmitter/receiver combination make up a transponder. Several transponders may share a single antenna. The total one-way transponder/antenna package is known as a repeater. As the name implies, the purpose of a repeater is to relay, or "repeat", a message with the highest possible fidelity. Two-way communications require at least two repeaters.

In a basic satellite repeater, signals are received and transmitted on separate frequencies to avoid mutual interference. The transmitted frequency is usually the lower of the two in order to simplify the onboard high-power electronic circuitry. The incoming carrier frequency is translated by a down-converter forming an integral part of the receiver. Down-conversion is achieved by mixing the received carrier with the waveform from an onboard crystal oscillator such that a harmonic beat equal to the difference between the two frequencies is produced. For example, when a carrier received at 6 GHz is mixed with a local oscillator waveform at 10 GHz, the net result is a 4-GHz carrier suitable for retransmission to the ground. The shape of the baseband signal (i.e. the intelligence riding on the carrier) does not change in the process.

In many applications it is desirable to manipulate the baseband signal during its passage from the satellite receiver to the transmitter. It may, for instance, be necessary to filter out, demodulate, amplify or re-route individual baseband signals arriving in some form of multiplex on the received carrier. At high frequencies (above 3 GHz), however, electronic components behave in a way which degrades the efficiency of signal manipulation. The solution lies in creating an intermediate frequency (IF) section between the receiver and the transmitter where a carrier frequency regime of only a few hundred MHz prevails. The IF is generated by down-converting the received carrier to a frequency which is much lower than the receive and transmit frequencies. After manipulation, the signal is upconverted and transmitted to the ground.

A more detailed description of receivers, transmitters and antennae follows.

Fig. 12.6. Typical satellite communications link (see example in text).

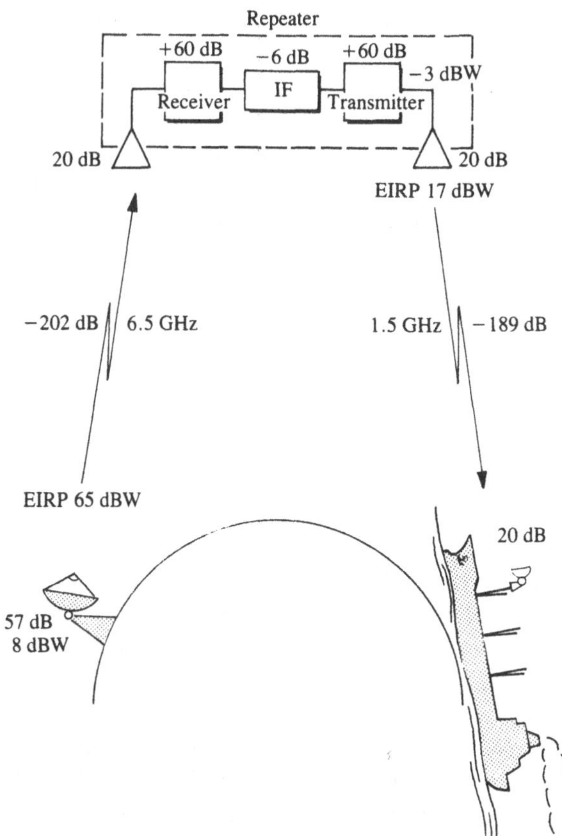

Receivers

A satellite communications receiver is a solid state device which usually consists of a low-noise amplifier (LNA) and a down-converter.

All electronic components with a resistive element generate thermal noise which tends to pollute the purity of the signal. Amplifiers are particularly prone to noise pollution. It can be demonstrated that the noise introduced by the receiver has a far more detrimental effect on the overall signal quality than any other unit downstream in the transponder; hence the need for particular care to build a "pure" receiver amplifier.

In the early days of spacecraft technology, bipolar transistors were employed in LNAs. Their noise temperature contribution grew rapidly from 100 K at 0.4 GHz to 1000 K at 6 GHz. As space communications moved to higher frequencies (11–30 GHz), the bipolar transistors gave way to tunnel diodes which added a fairly constant thermal noise level of 500 K to the signal. Nowadays, field-effect transistor (FET) technology is used in most LNAs. Their noise contribution ranges from 100 K at 1.5 GHz to 200 K at 14 GHz.

The down-converter comprises a stable, carefully tuned crystal oscillator and a frequency mixer. An undistorted sinusoidal waveform is obtained by filtering the crystal output. The mixer multiplies the crystal-generated waveform with the signal carrier to produce a harmonic beat. The beat constitutes the down-converted carrier.

Transmitters

The central feature of a satellite transmitter is the high-power amplifier (HPA). Depending on the particular application, the transmitter might also include pre-amplifiers ("drivers"), power dividers, power combiners, and power-limiting devices.

There are two types of HPA: transistorized power amplifiers (TPAs) and travelling wave tube amplifiers (TWTAs). TPAs belong to Class A, B or C depending on whether linearity (Class A and B) is more important than efficiency (Class B and C). TWTAs are more common than TPAs in satellite communications and merit a more detailed description.

A cross-section of a typical travelling wave tube is shown in Fig. 12.7. An electron gun serves the function of a cathode by generating a beam of electrons which are accelerated by an anode before entering a *slow-wave structure*. The slow-wave structure consists of a helix-shaped RF conductor and a set of beam-focussing permanent magnets. The coil and the magnets form a tunnel through which the electron beam can pass.

As the electron beam enters the tunnel, it begins to interact with the weak RF signal being fed to the input of the helix. The interaction causes a transfer of energy from the beam to the coil and produces a highly amplified RF signal at the output end of the coil. Meanwhile the spent electrons are picked up by a collector to prevent them from flowing back towards the anode.

The amplification characteristics of a TWTA is shown in Fig. 12.8. As the input signal level increases, the gain curve becomes non-linear. TWT saturation occurs when an increase in input produces no increase in output. Operating the tube in the non-linear region causes signal distortion. It is therefore customary to establish a working point on the curve with sufficient *back-off* from saturation to ensure tube operation in the linear region.

The TWTA is composed of the TWT itself and its electronic power conditioner (EPC) which transforms the 28–45-V bus voltage to the thousands of volts required by the TWT. The power consumption of TWTAs as well as TPAs is proportional to the RF input power. The ability of a transponder to accommodate RF signals is therefore said to be *power limited*, in addition to being frequency *bandwidth limited*. As explained earlier, a fundamental task of a communications system planner is to ensure that the two limitations are matched in an optimum fashion.

Fig. 12.7. Block schematic of travelling wave tube.

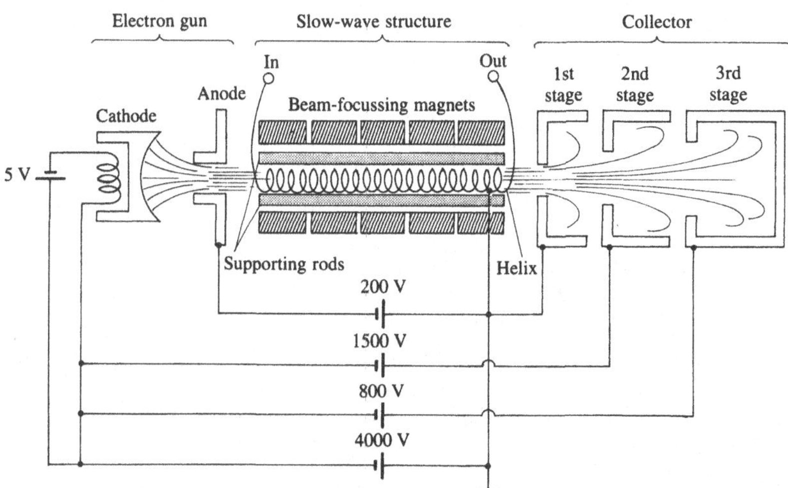

Antennae

The stable, earth-oriented attitude of satellites in geostationary orbit allows communications antennae to be highly directional, thereby focussing all the available RF power on the earth's disc. The earth subtends 17.4° from geostationary altitude. Global beam antennae cover the entire disc (Fig. 12.9), while shaped beam and spot-beam antennae (Fig. 12.10) provide localized geographical coverage (Fig. 12.11). The narrower the beam, the higher the antenna gain and the larger the diameter of the antenna aperture. It may seem paradoxical that larger antennae should give narrower beams, but the effect is the same as when you cup your hands behind your ear or around your mouth to improve the directivity of sound. For a given gain, the size of an antenna is proportional to the wavelength of the RF carrier, or inversely proportional to the frequency.

The spacecraft designer has a choice between several different types of antenna to suit particular applications. Helix antennae (Fig. 12.12) were common in the 1960s and 1970s at frequencies below 3 GHz. They were relatively rugged and provided adequate gain for global earth coverage. Horns (Figs 12.10 and 12.12) took over at higher frequencies. Nowadays, parabolic reflectors (Figs 12.10 and 12.13) predominate at

Fig. 12.8. Power transfer characteristics of a travelling wave tube amplifier.

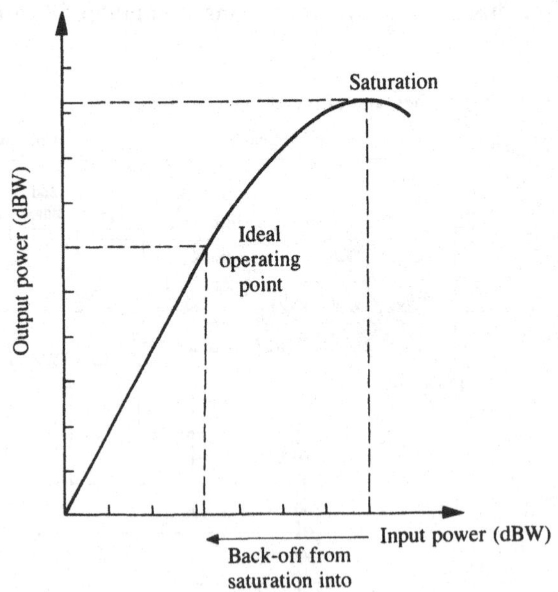

most communications frequencies above 1.5 GHz. By modifying the geometry of the reflector dish and the position of the feed, it is relatively easy to shape the radiation pattern into narrow beams with arbitrary contours.

Arrays of antenna elements (Fig. 12.14) or feeds are making inroads in applications where shaping of radiation patterns is desirable. Beam shaping is achieved by varying the amplitude and phase characteristics of individual elements across the array. Steered arrays change amplitude and phase dynamically for beam scanning or re-pointing. The electronically de-spun antenna is a form of steered array and

Fig. 12.9. Satellite antenna with global coverage.

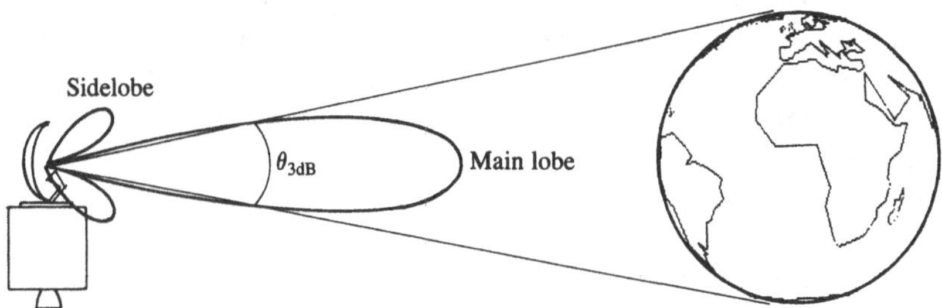

Fig. 12.10. INTELSAT-V satellite antennae.

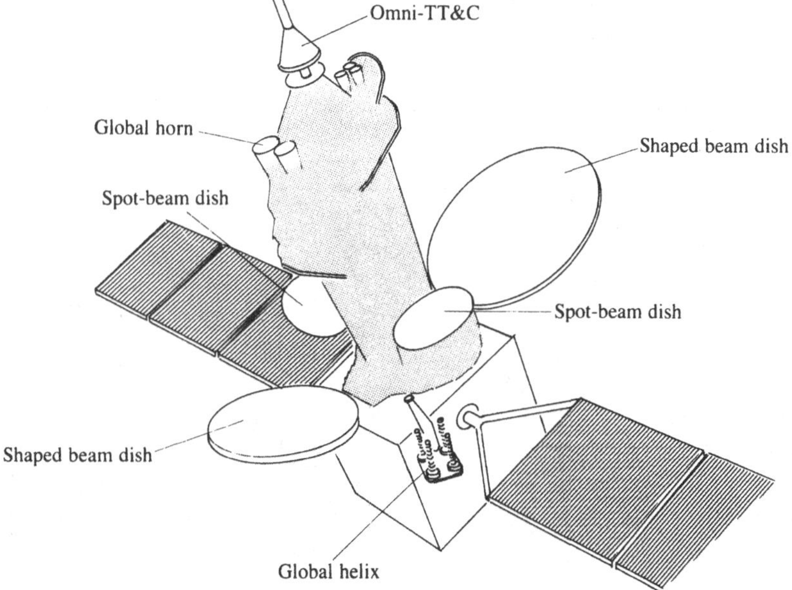

consists of dipoles mounted on a drum (Fig. 12.15). As the satellite spins, an electronic switching matrix inside the drum activitates and deactivates the dipoles in a cyclic, counter-rotating manner such that the electrical equivalent of a mechanically de-spun antenna is achieved. The toroidal pattern antenna (Fig. 12.16) offers a simple and reliable method for providing full-earth coverage, though RF power is wasted by the 360° radiation pattern.

The above inventory of antenna types is by no means exhaustive. As the difficulties grow in finding vacant frequency allocations, new antenna configurations are developed to enable *frequency re-use*. There are essentially two ways in which a repeater can make multiple use of a given frequency without causing radio interference onboard. One is to achieve spatial discrimination by avoiding overlap between adjacent receive and transmit spot-beams. The other method is to employ polarization discrimination whereby signals are received in one polarization sense and retransmitted in the opposite sense by the same antenna.

Link Budget

The transmission path from the ground to the satellite is called the *uplink*; the reverse path constitutes the *downlink*.

Although ground station transmitters often operate at power levels of several kW, the uplinked signal is extremely weak when it arrives at the

Fig. 12.11. Shaped antenna beam and spot-beam footprints illuminating major francophone regions from a satellite (Telecom-1) positioned over the Atlantic Ocean. Source: CNES.

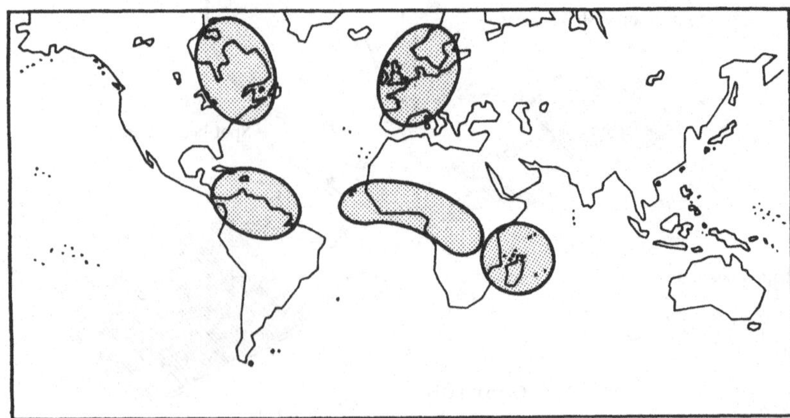

satellite receiver. In the course of its 40 000 km journey, its power flux density (PFD) decreases with the square of the distance. Further attenuation is caused by adverse weather conditions, polarization mismatch, antenna depointing and circuit losses. Limitations on available electric power onboard the satellite make the downlinked signal even weaker than the uplink by the time it is picked up by a receiving ground station. The gain of antennae, transmitters and receivers must therefore be carefully balanced if a communications mission is to succeed. The gain is defined as the ratio between the output power and the input power of an amplifying device. Transponder gain is a measure of active signal amplification onboard the spacecraft. Antenna gain is defined as the level of passive amplification of

Fig. 12.12. MARISAT satellite antennae. Source: COMSAT.

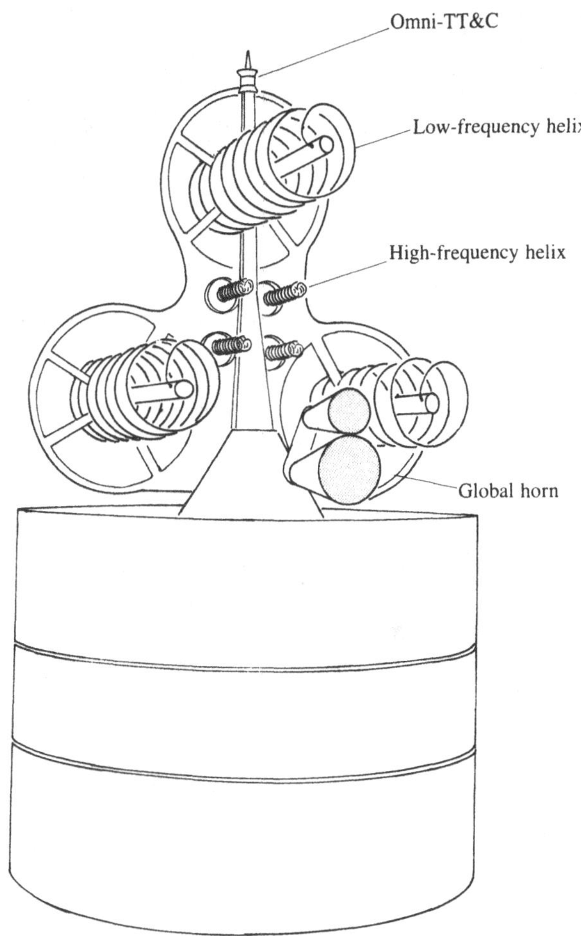

radiation in a preferred direction as compared to a perfectly isotropic, or spherical, radiation pattern. The higher the antenna gain, the lower the transmit power needed to achieve a specified power flux density in the preferred direction. The transmit power multiplied by the antenna gain is called the *equivalent isotropic radiated power* (EIRP). Expressed differently, the EIRP is the power which a perfectly isotropic antenna would have to transmit to achieve the same power flux density in all directions as a directional antenna in the preferred direction. This may

Fig. 12.13. MARECS satellite antennae. Courtesy British Aerospace.

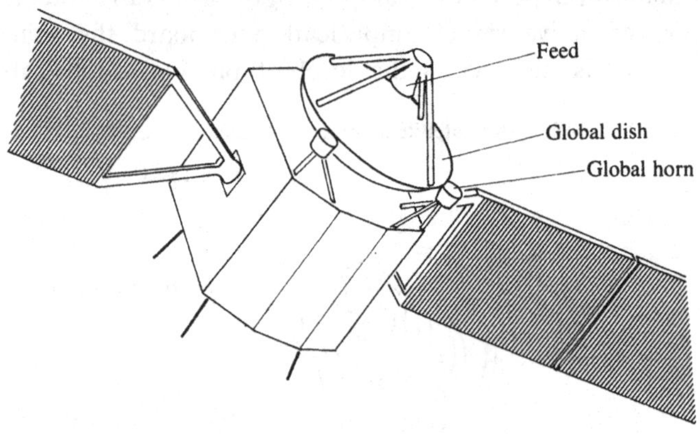

Fig. 12.14. INMARSAT-2 satellite antennae. Courtesy British Aerospace and INMARSAT.

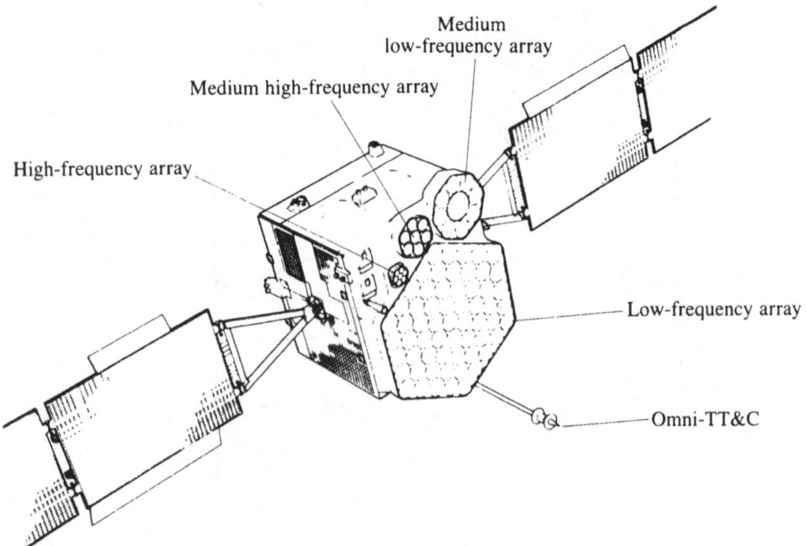

appear to be a needlessly convoluted parameter, but it does help in dissociating transmit power from antenna gain when optimizing communications systems.

Figure 12.6 illustrates how signal power changes along a ground-satellite-ground path due to losses and gains along the way.

Establishing a *link budget* between the satellite and the ground is the first step when attempting to optimize a space communications system. The following mathematical relationships should assist the reader in understanding how the various components in an end-to-end signal path affect the overall communications quality.

Antenna gain:

$$G = \eta A \frac{4\pi}{\lambda^2} = \eta \left(\frac{\pi D}{\lambda}\right)^2. \tag{12.8}$$

Antenna beamwidth:

$$\theta_{3\mathrm{dB}} = 70\lambda/D \qquad \text{(deg)}. \tag{12.9}$$

EIRP:

$$\mathrm{EIRP} = P_{\mathrm{t}}G_{\mathrm{t}} \qquad \text{(W)}. \tag{12.10}$$

Distance of spacecraft to station:

$$d = \sqrt{18.1839 - 5.3725\cos\phi}\ 10^7 \qquad \text{(m)} \tag{12.11}$$

$$\text{with } \phi \simeq -0.9\,\delta + 81 \qquad \text{(deg)}.$$

Fig. 12.15. METEOSAT satellite antennae. Source: ESA.

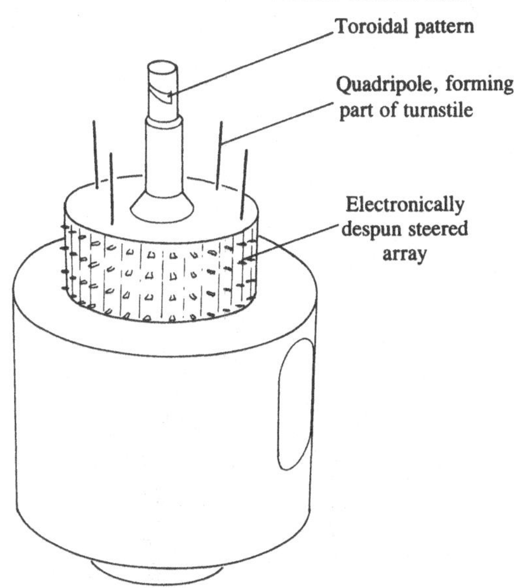

Power flux density:

$$\text{PFD} = \text{EIRP}/(4\pi d^2) \qquad (\text{W/m}^2). \tag{12.12}$$

Received power:

$$P_r = \text{PFD}\,\eta A = \tag{12.13}$$

$$= \text{EIRP}\,G_r \frac{1}{L}\left(\frac{\lambda}{4\pi d}\right)^2 \qquad (\text{W}). \tag{12.14}$$

FM modulation index:

$$\beta = \Delta f / f_{\text{max}}. \tag{12.15}$$

FM bandwidth:

$$B = 2 f_{\text{max}}(1 + \beta) \qquad (\text{Hz}). \tag{12.16}$$

Noise power:

$$N = kTB \qquad (\text{W}). \tag{12.17}$$

Carrier-to-noise ratio:

$$\frac{C}{N} = \frac{P_r}{N} = \frac{\text{EIRP}}{kB} \cdot \frac{G_r}{T} \cdot \frac{1}{L}\left(\frac{\lambda}{4\pi d}\right)^2, \tag{12.18}$$

where G_r/T = receiver figure of merit.

Total system C/N:

$$(C/N)_t^{-1} = (C/N)_u^{-1} + (C/N)_d^{-1}. \tag{12.19}$$

Signal-to-noise ratio:

$$S/N = 3\beta^2(\beta + 1) \cdot C/N. \tag{12.20}$$

(assuming FM modulation by pure sinusoid),

Fig. 12.16. Cross-section of a toroidal pattern antenna beam.

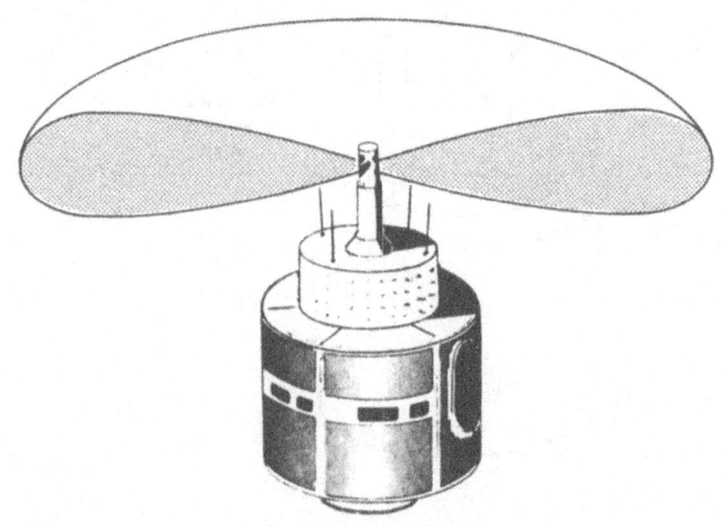

where η = antenna efficiency (typically 0.65),
A = area of antenna aperture (m^2),
D = diameter of antenna aperture (m),
P_t = transmitted power (W),
P_r = received power (W),
G_t = transmit antenna gain,
G_r = receive antenna gain,
d = distance between ground station and satellite (m),
ϕ = geocentric angle of ground station (deg),
δ = satellite elevation above local horizon (deg),
λ = wavelength (m),
Δf = maximum frequency deviation from carrier frequency in FM (Hz),
f_{max} = maximum modulating frequency (Hz),
k = Boltzmann's constant (1.38×10^{-23} J/K),
T = noise temperature at receiver input (K),
L = loss factors due to weather conditions, depointing, polarization mismatch, circuit losses,
$(C/N)_u$ = uplink C/N, and
$(C/N)_d$ = downlink C/N.

Example

You are designing a communications system for transmitting a voice signal (0–3600 Hz) from a ground station via a geostationary communications satellite to a small receiving station aboard a ship at sea. Assume that the two stations are located at the extreme edge of satellite visibility, such that the satellite is seen from the stations at 0° elevation above the local horizon (Fig. 12.6). The signal is uplinked in FM on a 6.5-GHz carrier and is downlinked at 1.5 GHz. The maximum frequency deviation Δf is 11 400 Hz. The noise temperature at the satellite receiver is 1250 K, while the noise temperature at the ship station receiver is 225 K. All the antennae in the system are parabolic with an efficiency of 0.65. The diameter of the ground transmit antenna is 13 m, that of the satellite receive antenna 0.185 m, that of the satellite transmit antenna 0.8 m, and that of the ship receive antenna also 0.8 m. Find the voice quality in terms of overall S/N at the receiving ship station.

Table 12.1 suggests an answer based on Eqns 12.8–12.20. Note that the results are given in two columns, the first in linear values and the second

Table 12.1. *Simplified link budget for a maritime satellite voice communications channel*

Constants	Uplink		Downlink	
Frequency	6.5 GHz		1.5 GHz	
Wavelength	4.6 cm		20 cm	
Transmit antenna diameter	13 m		0.8 m	
Receive antenna diameter	0.185 m		0.8 m	
Antenna efficiency	0.65		0.65	
Modulating frequency	3.6 kHz		3.6 kHz	
FM frequency deviation	11.4 kHz		11.4 kHz	
Modulation index	3.167		3.167	
Distance ground to satellite	43 332 270 m		43 332 270 m	
	Linear	dB	Linear	dB
Transmitted power	6.3 W	8 dBW	0.5 W	−3 dBW
Transmit antenna gain	508 960	57.1 dB	102.6	20.1 dBW
EIRP	3 206 450 W	65.1 dBW	51.3 W	17.1 dBW
Free space loss	7.18×10^{-21}	−201.4 dBW	1.35×10^{-19}	−188.7 dB
Other losses	1.175	0.7 dBW	1.38	1.4 dB
Boltzmann's constant	1.38×10^{-23} J/K	−228.6 dBW/K/Hz	1.38×10^{-23} J/K	−228.6 dBW/K/Hz
Temperature	1250 K	31 dBK	225 K	23.5 dBK
Bandwith	30 kHz	44.8 dBHz	30 kHz	44.8 dBHz
Receive antenna gain	103	20.1 dB	103	20.1 dB
Received power	2×10^{-12} W	−116.9 dBW	5×10^{-16} W	−152.9 dBW
G/T	0.082 K^{-1}	−10.8 dB/K	0.456 K^{-1}	−3.4 dB/K
C/N	3905	35.9 dB	5.5	7.4 dB
System *C/N*			7.4 dB	
System *S/N*			28.4 dB	

in logarithms. The logarithmic value *Lo* is obtained from the linear figure *Li* through the relationship

$$Lo = 10 \log (Li) \qquad \text{(dB)}.$$

Logarithms are of course dimensionless but, depending on the unit of *Li*, it is customary to write the same unit after "dB" so as to keep track of the origin of the corresponding *Lo*. For example, if *Li* is in watts, we would express *Lo* in "dBW", and if *Li* is in Kelvin, *Lo* becomes "dBK", etc.

Logarithms are preferred to linear values in link budget calculations for two reasons: they make the often unwieldy linear figures more manageable, and they reduce multiplications and divisions to simple additions and subtractions. Note that a 3-dB increase in a particular parameter is the same as doubling its value, since $3 \text{ dB} \simeq 10 \log 2$. Similarly, a decrease of 3 dB means halving the value ($-3 \text{ dB} \simeq 10 \log 1/2$). This explains, for instance, why the half-power antenna beamwidth is denoted $\theta_{3\,\text{dB}}$, signifying that the useful beam ends where the antenna gain is down by 3 dB from its peak value (Fig. 12.9).

The link budget calculations shown in Table 12.1 are not sufficiently detailed for procuring satellite and ground station communications hardware. In a real communications system, analogue and digital baseband signals are often *multiplexed* (mixed) on a single carrier, or several different carriers may be relayed simultaneously through the satellite transponders (*multiple access*, see above), such that the bandwidth requirements and the signal quality finish up being different from the above example.

Bibliography

Bleazard, G. B. (1985). *Introducing Satellite Communications*. Manchester: NCC Publications.
Fontolliet, P. G. (1986). *Telecommunication Systems*. Dedham, USA: Artech House.
Fthenakis, E. (1984). *Manual of Satellite Communications*. New York: McGraw-Hill.
Maral, G. and Bousquet, M. (1986). *Satellite Communications Systems*. Chichester: John Wiley.
Miya, *et al.* (1981). Satellite Communications Technology. Tokyo: KDD Engineering and Consulting.

13

Meteorological Payload

Introduction

Atmospheric pressure, temperature, humidity, wind speed, and sea surface temperature are fundamental parameters for determining the weather. If it were possible to measure these five parameters simultaneously across the globe at regular time, distance and height intervals, the problems of medium- and long-range weather forecasting would be greatly reduced.

Before the advent of satellites, global weather measurements on such a scale would have been totally impracticable. Systematic meteorological observations used to be extremely sparse. Although most industrialized countries maintained a network of ground-level observation stations, vertical sounding was limited to launching occasional balloons from land and sea, and receiving sporadic reports from commercial aircraft. These horizontal and vertical measurements covered only a small portion of the earth, forcing meteorologists to bridge the gap with educated guesses.

In the late 1940s, sounding rockets equipped with cameras took pictures of the earth from altitudes of 100 km and more. These early photographs revealed a whole new family of physical relationships in the atmosphere. In the years that followed the launch of Sputnik in 1957, much effort was devoted to developing television cameras and radiometers suitable for satellite meteorology. Throughout the 1960s, the United States and the Soviet Union deployed a host of increasingly powerful weather satellites which transmitted visible and infrared images to ground stations on the earth. The first steps towards making global weather observations had been taken.

Low-orbiting Satellites

The early meteorological satellites were launched into low earth orbit by necessity, since rockets in those days had limited lifting capacity. The satellites were placed in highly inclined orbits to provide global coverage over the span of a few days. The field of view of their cameras cut 100–1000 km swaths across the globe with a picture resolution of a few kilometres. The visible (VIS) pictures showed the distribution of clouds, while the infrared (IR) images measured cloud top temperatures so that a first estimate of cloud height and sea surface temperature could be made. This was still a long way from measuring pressure, temperature, humidity and wind speed, but it was at least a step in the right direction.

The main advantages of low-orbiters are the global coverage and the achievable picture resolution. For these reasons some meteorological satellites continue to be launched into low, highly inclined orbits. A major disadvantage, however, is the lack of spatial and temporal imaging continuity. It is difficult to follow the evolution of a weather pattern if the time between satellite passes is so long that the pattern is significantly different from one image to the next. One solution is to launch more satellites and stagger their orbital nodes, but because of the high cost of satellites and their relatively short lifetime, this solution has obvious limitations.

Geostationary Satellites

The ultimate solution to the temporal continuity problem lay in launching geostationary weather satellites. In the course of the 1970s, the USA, Europe, Japan and India launched a family of such satellites, spacing them more or less evenly along the equator (Fig. 13.1) in order to achieve near-global coverage. Given that each satellite covers almost half the globe, one might have argued that two satellites 180° apart would have sufficed for full-earth coverage. The curvature of the earth, however, renders the peripheral parts of an image unusable, and additional satellites are therefore needed to obtain reasonable spatial continuity around the globe. The same curvature degrades the usefulness of images in high-latitude regions, which is why countries in the extreme north and south have always taken a rather lukewarm interest in geostationary meteorological satellites.

The geostationary orbit has to be almost perfectly circular and equatorial if successive images are to be comparable. A slightly inclined orbit would cause the earth scene to wobble in a north–south direction as the satellite describes its characteristic figure "8" sub-satellite path (see Fig. 3.1), and orbital eccentricity would produce an unacceptable east–west libration.

Subsystem Architecture

Strictly speaking, a geostationary weather satellite possesses several different payloads, not just one. The primary payload is the radiometer which measures earth radiances in the *visible* (0.4–1.0 μm) and *thermal infrared* (10–13 μm) spectral bands. Some satellites carry an additional radiometer channel which operates in the *water-vapour* absorption band (5.5–7.5 μm). The secondary payload is the *communications transponder* for transmitting "raw" images to a central ground processing facility, and for disseminating processed images from the central facility to smaller user stations scattered across the coverage area. The satellites also accommodate additional transponders to relay environmental observations from small, inexpensive and unmanned *data collection platforms* deployed in inaccessible regions on land, as well as onboard ships and aircraft. Figure 13.2 shows the main components of meteorological satellite payloads.

The Radiometer

A typical radiometer of the kind used onboard geostationary

Fig. 13.1. Full-earth coverage of five meteorological satellites. The 30° elevation contours constitute limits for useful imagery, while the 0° contours show the geographical limits for useful communications.

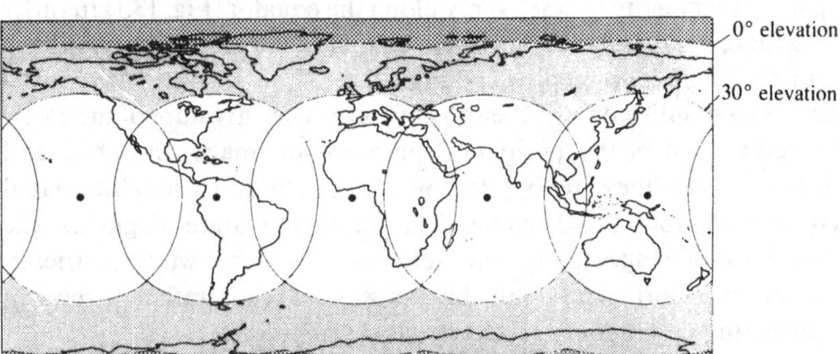

satellites is shown in Fig. 13.3. It consists of a mirror telescope, a step motor, a set of radiance detectors, and an IR detector cooler.

A radiometer telescope mounted on a spin-stabilized satellite scans the earth's disc from east to west by virtue of the satellite's rotation. Using the European satellite METEOSAT as an example, each detector output is sampled 2500 times in the course of a scan. A sample takes the form of an 8-bit digital data word which quantifies the radiation level of that particular picture element, or *pixel*.

Given that the earth subtends 17.4°, an earth image line has to be collected by the radiometer during 1/20 of the satellite spin period. By buffering the data onboard before transmission, it is possible to "stretch" the scan line over the remaining 19/20 of the spin period so that the transmission data rate, and hence the required transmitter bandwidth and power, are reduced.

The first scan line is recorded while the telescope is aligned with the southern rim of the earth's disc. A line consists of 2500 pixels. After each scan, the telescope is pivoted a small step towards the north, using the sun or earth sensor output as the prompt. After 25 min, the telescope has achieved 2500 steps and is now positioned at the northern rim. The 2500 east–west samples in combination with 2500 south–north steps provide a 2500 × 2500 raster of pixels covering the earth's entire disc (Fig. 13.4). Since the earth's diameter is 12 742 km, the image *resolution* at the sub-satellite point becomes 12 742/2500 = 5 km.

When the scan is completed, the telescope needs 2.5 min to retrace to the southern rim. The rapid motion gives rise to spacecraft nutation

Fig. 13.2. Typical meteorological satellite payload architecture.

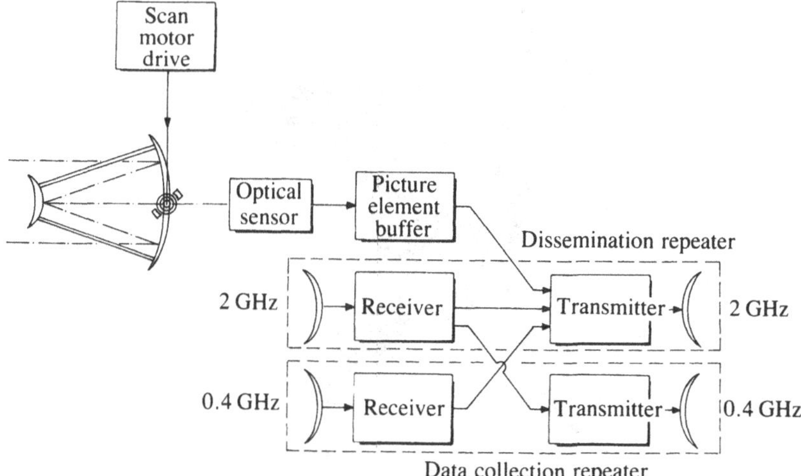

which is dampened out during the remaining 2.5 min before the next picture scan begins. It thus takes exactly 30 min to generate an image – or, in the case of METEOSAT, three images, since the satellite is equipped with three detectors operating at different spectral bands.

The most notable components of a radiometer telescope are the primary and secondary optical reflectors. The primary reflector has a paraboloid shape, while the secondary reflector has a spherical surface. Their interaction is analogous to that of a Cassegrain antenna for RF signal reception, i.e. the primary reflector projects incoming parallel radiation in the secondary reflector which, in turn, focusses the radiation energy into a feed (Fig. 13.5).

The "feed" of the radiometer telescope is a pair of folding mirrors aligned with the pivot axis of the telescope, such that the incoming radiation is deflected towards the fixed sensor package regardless of the angular position of the telescope. The sensor package comprises a focussing mirror, a VIS/IR spectral separation mirror, and the VIS and IR detectors. The VIS detectors are silicon photo diodes, whereas the IR detectors employ mercury cadmium telluride as the detecting medium.

Fig. 13.3. METEOSAT radiometer. Source: ESA.

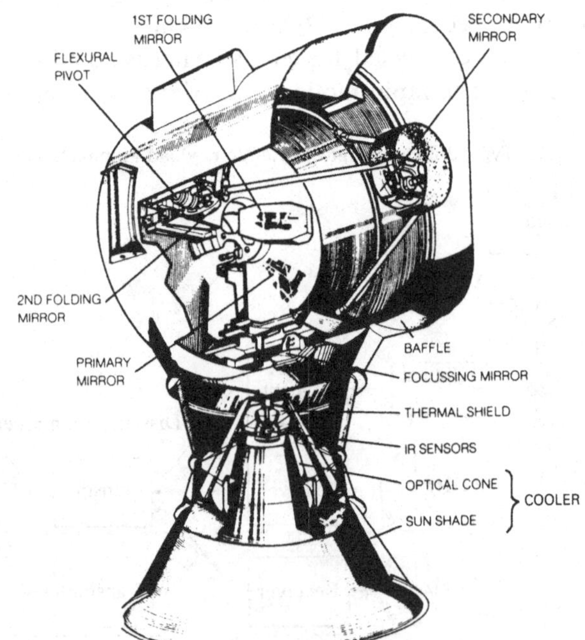

Fig. 13.4. Pixel composite of earth image.

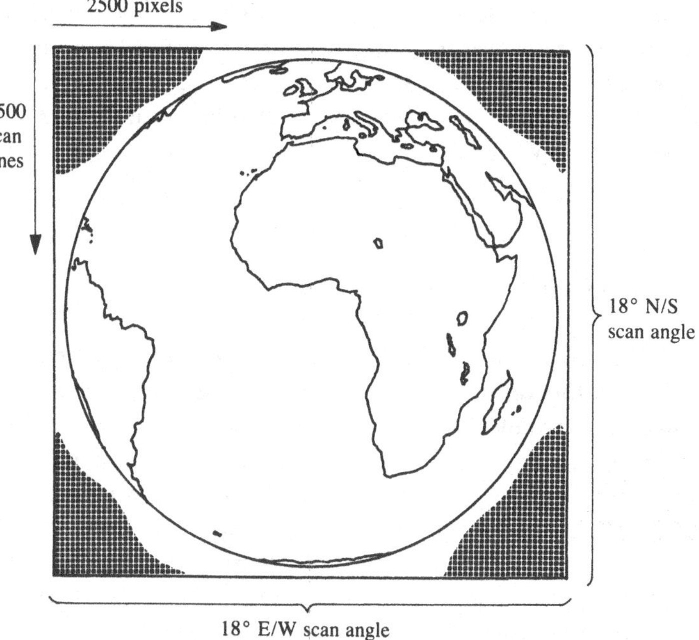

Fig. 13.5. Details of METEOSAT radiometer optics. Source: ESA.

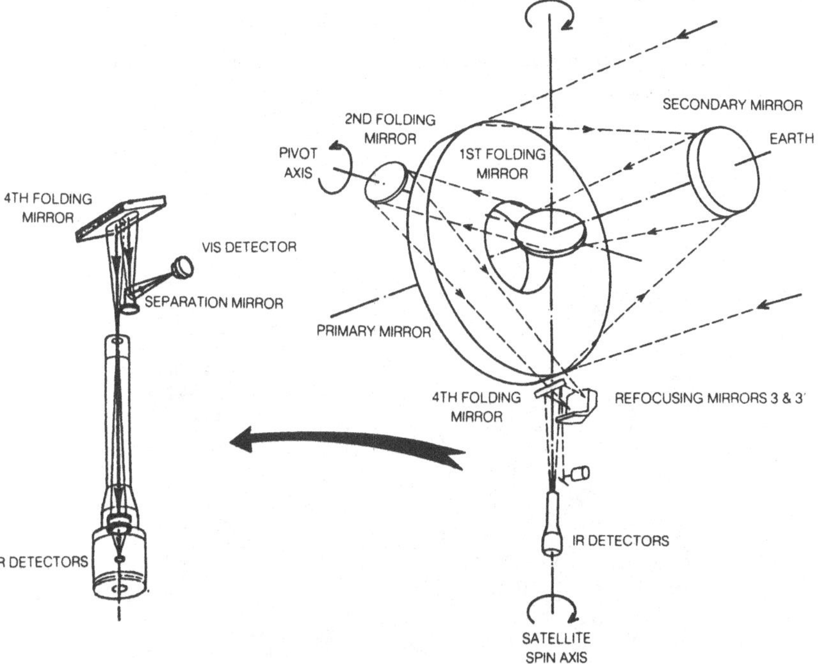

Since IR detectors are basically temperature sensors, it is essential that they be kept as cold as possible, otherwise the ambient temperature would predominate over the faint IR radiance received from the earth. The IR sensors are therefore mounted at the apex of a conical "chimney", or *cooler*, which looks straight into space (Fig. 13.3). This arrangement allows the detectors to be maintained at temperatures of about 80 K.

Meteorological Data Extraction

Interpretation of satellite imagery is made by computers and human operators on the ground. Wind speed is readily obtainable by tracking the movement of clouds from one VIS image to the next (Fig. 13.6). Since clouds get colder with altitude, the thermal IR images provide an indication of cloud height (Fig. 13.7), allowing wind speeds at different layers of the atmosphere to be determined. The same images measure sea surface temperatures which influence the evolution of weather patterns. Air pressure can, to some extent, be inferred from wind speed and direction. Images from the water vapour channel (Fig. 13.8) are difficult to interpret in terms of humidity due to the poor

Fig. 13.6. METEOSAT image in the visible spectrum. Photo ESA.

Fig. 13.7. METEOSAT image in the infrared spectrum. Photo ESA.

Fig. 13.8. METEOSAT image in the water-vapour spectrum. Photo ESA.

altitude resolution, but they do provide information about air movements even in areas where there are no clouds. In addition to serving weather forecasters, all three spectral channels provide invaluable information to explain a wide variety of physical and dynamic phenomena in the atmosphere.

Bibliography

Introduction to the Meteosat System (1981). ESA SP-1041. Paris: European Space Agency.

Neiburger, M., Edinger, J. G. and Bonner, W. D. (1973). *Understanding Our Atmospheric Environment*. San Francisco: W. H. Freeman.

14

Product Assurance

Introduction

Geostationary applications satellites are expected to function in orbit for typically 7–10 years. This is not a trivial requirement when considering the debilitating influence of their environment (Chapter 4). Nor can they be repaired in orbit. It is true that in the heyday of the space shuttle some low-orbiting spacecraft were retrieved and repaired, but satellites in geostationary orbit will be out of reach for decades to come.

The customer usually specifies the orbital lifetime requirement in terms of mission survival for a number of years Y with a probability P. "Mission survival" means that the payload function should be operating to specification, even if there are equipment degradations or failures in the spacecraft. Typically:

$$P(Y \geqslant 10 \text{ years}) = 0.75.$$

Thus formulated, the customer acknowledges the risk that there may be a 25% chance of mission failure in less than 10 years. More importantly, however, he requires the contractor to demonstrate through analysis that the design is good for 10 years or more with a 75% probability.

The contractor has several ways of complying. He can use space-qualified components and materials whose failure modes and degradation characteristics have been established through ground testing and flight experience, or he may duplicate components or units to achieve redundancy. By optimizing their configuration, he should be able to extend their combined lifetime to the maximum.

The mission lifetime objective is achieved by exercising an engineering discipline called *product assurance*. The main sub-disciplines are component selection, materials and processes selection, reliability assurance and quality assurance.

Component Selection

A component is an electrical, electronic or electromechanical device which performs a single function and is made of elements which cannot be separated without destroying the function. Examples are resistors, capacitors, inductors, transistors, transformers, crystals and filters. The words "component" and "part" are used interchangeably in the space trade.

The major space organizations have established *preferred parts lists* (PPL). The lists provide design and peformance data about components which have been successfully qualified on the ground or in flight. As long as the contractor selects his components from a list approved by the customer, there will be no basic disagreement between the two. Even so, the contractor has to ascertain that the particular batch, or lot, of approved parts which he is procuring satisfies a set of quality criteria. This he does through screening of samples from the lot. Screening procedures vary depending on the type of component, but most are subjected to "pre-cap" inspection (i.e. examination before encapsulation), thermal cycling, acceleration, vibration, seal tests, X-ray, and destructive physical analysis (DPA). The whole lot is rejected if a predetermined number of components fail the screening tests.

In practice, a contractor is seldom able to satisfy all his component needs from the PPLs, e.g. because:

(i) the customer does not recognize the validity of a particular list;

(ii) the manufacturing of an approved part is discontinued; and

(iii) new parts come on the market which are potentially superior to existing components but which have not yet been space qualified.

While problem (i) can often be solved through persuasion, item (ii) requires the contractor either to find an alternative approved part or to change his equipment design. In the case of (iii) he is obliged to submit the new parts to qualification testing according to procedures which in turn must be approved by the customer.

Components are seldom developed for space applications alone. Most spacecraft parts stem from commercial production lines but undergo particularly rigorous quality control before delivery. The manufacturer's financial incentive to produce space-qualified components is relatively minor, and his dedication to meeting quality requirements and delivery deadlines suffers as a result. Parts procurement is a tortuous process for the satellite contractor who often has to alter his design or his schedule to match the changing component market.

Materials and Processes

Metals, plastics, adhesives, paints, varnishes and lubricants are examples of materials which can create problems in space applications. When exposed to cosmic radiation, adhesives, paints and varnishes may become brittle or change colour to the point where they can no longer serve their intended purpose. Lubricants are prone to decompose in vacuum, causing moving parts to seize up while contaminating sensitive surfaces such as optical lenses.

Materials must be selected with the same care as components. There are lists of approved materials, and there are procedures stating how to space-qualify new materials.

A *process* is the method by which a material is manufactured or applied in spacecraft construction. Examples are welding, soldering, bonding, painting, metal finishing, crimping of wires and encapsulation. By following an approved and well-tried process, the spacecraft builder can minimize the risk of quality defects.

Reliability

The statistical failure rate $\lambda(t)$ of many components or units changes with time according to a "bathtub curve" (Fig. 14.1). There are three distinct phases of failure rate, labelled "infant mortality", "random failure" and "wear-out".

The infant mortality rate can be quite high. Those components which survive settle down to a fairly constant random failure rate over many years. Towards the end a wear-out mechanism sets in, resulting in a rapidly increasing failure rate.

Infant mortality can be largely eliminated through component or equipment *bake-out*. The hardware is placed in an oven and is baked for several days in order to provoke dominant failure modes in weak parts without adversely affecting the lifetime of good components. The process is designed to promote the survival of the fittest. As in the case of screening, the entire batch is rejected if the number of failed parts exceeds a specified threshold.

The detrimental effect of *random failures* can be reduced by implementing redundancy, i.e. by duplicating parts or units and connecting them in series or in parallel. Since the random failure rate is approximately constant (λ), it is possible to estimate a numerical value of λ for any one part based on test and flight experience. Some of the major space organizations have issued handbooks giving approximate λ values for every conceivable component. An example is given in Table 14.1. Here, λ is measured in number of *Failures In Ten-to-the-nine (10^9) hours* (FITs).

Not surprisingly, the random failure rate λ is a function of the extent to which a component is stressed. The family of curves in Fig. 14.2 is normally provided by the component manufacturer. It shows how the failure rate of an electric relay varies with the amount of current flowing through the contacts, as well as with temperature. In this particular

Table 14.1. *Typical component failure rates per 10^9 h at 20°C for components derated to 25% of nominal performance ceiling*

Type of component	FIT (λ = FIT$*10^{-9}$)
Resistors	5–200
Capacitors	3–100
Diodes	4–100
Filters	10–25
Couplers, circulators, switches	10–20
Bearings	35
Rotating joints	1000
Electric motors	200
Transistors	10–50
Travelling wave tubes	2000
Crystals	80
Relays	400

Fig. 14.1. "Bathtub" curve of component failure rates.

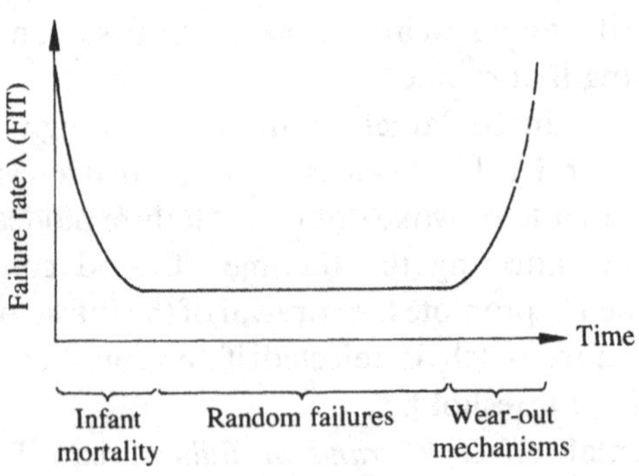

example, the 100% curve at 100°C is the maximum *rated* value of electric current and temperature; it is the maximum allowable stress according to the manufacturer. The spacecraft designer now has the option to reduce the relay's random failure rate by deliberately reducing the stress levels in his design. He is in effect *derating* the component to lower than nominal operating levels in order to increase its reliability.

Knowing λ, it is possible to estimate the probability $P(t)$ of a component failing within a certain time t (hours):

$$P(t) = \exp(-\lambda t). \tag{14.1}$$

For example, take a relay which is normally open, but which needs to close an electric circuit repeatedly with a certain reliability. With $\lambda = 400$ FITs it has a probability of surviving a 7-year satellite mission equal to:

$$\begin{aligned} P(t) &= P(7 \times 365 \times 24 \text{ h}) \\ &= \exp(-400 \cdot 10^{-9} \times 7 \times 365 \times 24), \\ P(t) &= 0.97577. \end{aligned}$$

This probability is a measure of the reliability of the relay.

Having maximized the reliability of components, the designer can now improve the reliability of units by judiciously duplicating components in the circuit layout. According to elementary theory, the

Fig. 14.2. Component failure rates as a function of de-rating.

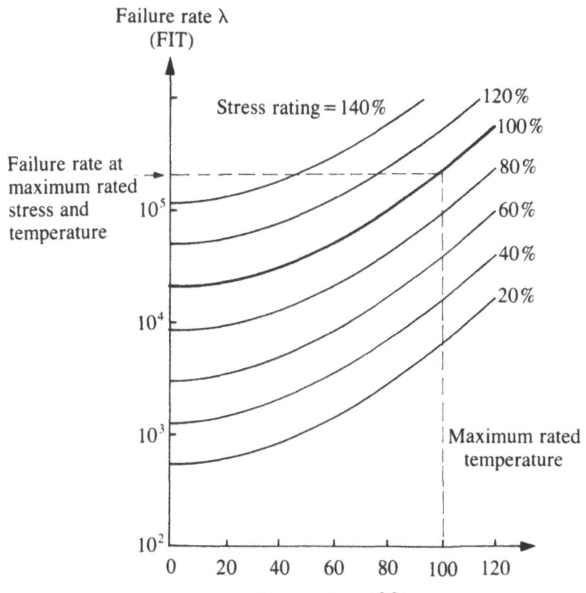

reliability P_{AB} of two components A and B in series (Fig. 14.3) equals:

$$P_{AB} = P_A P_B \tag{14.2}$$

where P_A and P_B are the reliability of each component. If the two components are connected in parallel instead of serially (Fig. 14.4), we have:

$$Q_{AB} = Q_A Q_B \tag{14.3}$$

with

$$Q_A = 1 - P_A, \qquad Q_B = 1 - P_B$$

whence:

$$P_{AB} = 1 - Q_{AB} = P_A + P_B - P_A P_B.$$

Let us again use relays to illustrate the point. If two identical relays A are connected in series, we find from Eqn 14.2 that the probability of the circuit being closed is:

$$P_{AA} = 0.97577 \times 0.97577 = 0.95213,$$

while according to Eqn 14.3, wiring the two relays in parallel gives us:

$$Q_{AA} = (1 - 0.97577)(1 - 0.97577) = 0.000587.$$

Since $P_{AA} = 1 - Q_{AA}$:

$$P_{AA} = 0.99941.$$

As one might have expected, two closing relays in series are *less* reliable than one relay by itself, whereas two relays in parallel are *more* reliabile. (The opposite would be true, however, if the primary concern were the ability of the relays to *open* rather than *close* the circuit. The probability-minded reader might try to prove this mathematically.)

Starting out from these simple rules, it is possible to calculate the reliability of circuits of arbitrary complexity, and hence of electrical and electronic units (Fig. 14.5). Similar techniques are used to analyse the

Fig. 14.3 (Left). Reliability diagram of two components in series. Fig. 14.4 (Right). Reliability diagram of two components in parallel.

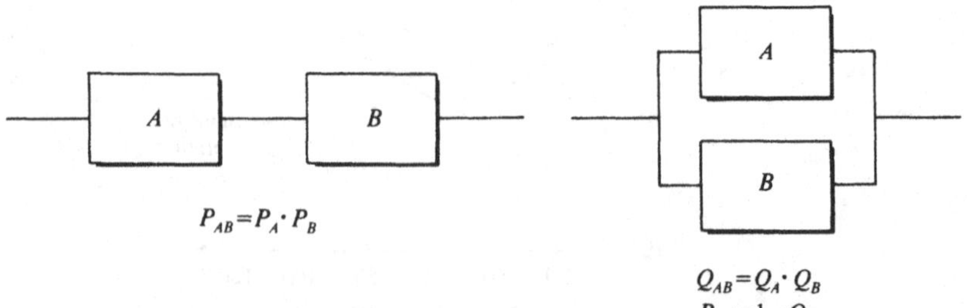

reliability of mechanical units, and ultimately of the entire spacecraft. Thus the spacecraft contractor is able to demonstrate, by analysis, that the satellite will survive Y years in orbit with a probability of $P(Y)$ before *wear-out mechanisms* cause it to fail for good.

Quality Assurance

Often abbreviated "QA", quality assurance is a technical surveillance activity aimed at systematically overseeing satellite hardware procurement, manufacturing, integration and test from the viewpoint of quality. The customer's and the prime contractor's QA engineers look over the shoulder of fellow engineers and technicians in the drawing office, in the laboratory and on the shop floor. Their inspection mandate extends to the premises of subcontractors as well as suppliers, and they are as active on the launch site as at home base. Should a QA engineer have reservations about quality during work in progress, he will inform his project manager. If the latter fails to take action, the QA engineer has privileged access to the next tier of management in order to be heard. Effective quality assurance is a prerequisite for a viable applications satellite programme.

Quality assurance is exercised through a number of well defined activities. Manufacturer audits are organized regularly to verify the adequacy of production environment, spacecraft equipment workmanship, and training of technical staff. A company found lacking during an audit could be barred from bidding on future space contracts. A great deal of QA effort is devoted to examining the completeness and consistency of project documentation, such as specifications, manufacturing drawings, test procedures and log books. The QA engineer's

Fig. 14.5. Composite reliability diagram.

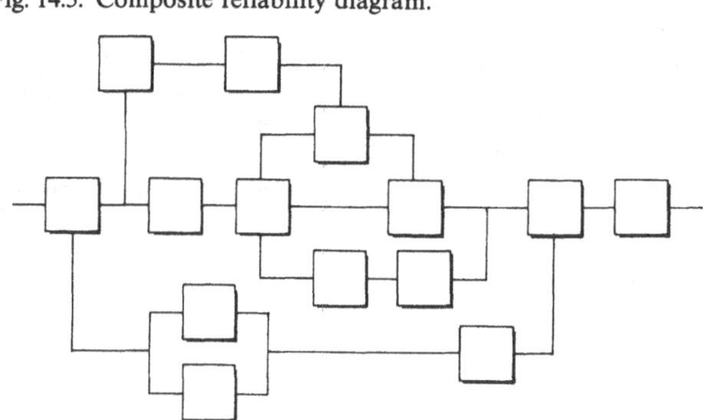

co-signature is required before quality-controlled documents may be released.

QA staff are key participants in the review of non-conformance and failures. A non-conformance exists whenever a spacecraft equipment performs below specified levels. If the non-conformance can already be predicted at the design stage, the manufacturer may raise a *waiver request* to the prime contractor who may agree or refuse to grant it. A refusal obliges the manufacturer to redesign until the specification is met. If the non-conformance is discovered only after the equipment has been built and tested, the manufacturer raises a *deviation request* which is dealt with in a similar manner to a waiver. Should the non-conformance violate overall mission performance specifications, the prime contractor is obliged to seek the ultimate customer's approval of a waiver or deviation request.

The failure of any piece of equipment during test is always regarded as a serious event. Attempts to find the cause of a failure are conducted under the auspices of a Material Review Board (MRB) which also issues instructions regarding remedial action.

The QA function also encompasses incoming inspection of delivered hardware, as well as surveillance of testing, handling, packaging, shipping and storage of spacecraft equipment.

Bibliography

Bazovsky, I. (1961). *Reliability Theory and Practice*. London: Prentice-Hall.

Basic Requirements for Product Assurance of ESA Spacecraft and Associated Equipment (1981). ESA PSS-01-0, Issue 1. Noordwijk: European Space Research and Technology Centre.

Product Assurance Management and Audit Systems for ESA Spacecraft and Associated Equipment (1981). ESA PSS-01-10, Issue 1. Noordwijk: European Space Research and Technology Centre.

Quality Assurance of ESA Spacecraft and Associated Equipment (1981). ESA PSS-01-20, Issue 1. Noordwijk: European Space Research and Technology Centre.

Reliability Assurance of ESA Spacecraft and Associated Equipment (1981). ESA PSS-01-30, Issue 1. Noordwijk: European Space Research and Technology Centre.

Component Selection, Procurement and Control for ESA Spacecraft and Associated Equipment (1981). ESA PSS-01-60, Issue 1. Noordwijk: European Space Research and Technology Centre.

Material and Process Selection and Control for ESA Spacecraft and Associated Equipment (1981). ESA PSS-01-70, Issue 1. Noordwijk: European Space Research and Technology Centre.

15

Spacecraft Development and Testing

Introduction

In an ideal world, spacecraft development would begin at the preliminary design stage and stop when the complete design is finalized, the hardware manufactured, and the satellite assembled. The next logical step would be to *test* the spacecraft in order to ensure compliance with specified performance, establish design and performance margins, and verify that the workmanship has been flawless. The test phase would end after the satellite has been checked out in orbit and is ready to go into service.

Alas, in the real world of spacecraft construction, the sequence of events is not so straightforward. In this chapter we will attempt to steer the reader through a labyrinth of phased development, hardware hierarchy and heritage, model philosophy, assembly, integration and test. For the spacecraft builder, the path is fraught with technical, financial and schedule pitfalls, and his management skills are constantly put to test. The customer, meanwhile, looks nervously over the shoulder of the builder to ensure that no effort is spared to produce a satellite which offers maximum value for money.

Spacecraft Development

Satellites which make use of new technology or particularly complex designs undergo a *phased development*. By allowing for planned interruptions at well-defined points during the development programme, it is possible to take stock and change the approach without jeopardizing the programme as a whole. Phase A includes mission definition and technical feasibility studies. Phase B covers system design, while Phase C encompasses detailed design finalization and prototype validation. During Phase D, flight hardware is built and tested. The precise definition of the phases varies somewhat from project to project. A typical phased development programme takes between 4 and 6 years.

In applications satellite programmes where timely delivery tends to be more important than using the latest designs and technology, the various development stages show considerable overlap, and any changes of mind along the way can become quite costly in terms of money and scheduling. The first step is for the customer to formulate his specifications. These are sent out in the form of a call for tender or request for proposals to potential bidders in industry who are invited to make competitive offers. The customer evaluates the offers received and selects the winning bidder on the basis of quality, price, and delivery time. After the contract is awarded, the winning bidder becomes the prime contractor who undertakes to build one or several satellites with the help of a consortium of subcontractors. Figure 15.1 shows a hypothetical industrial consortium contracted by a customer to build and test a meteorological satellite.

Traditionally, the customer has awarded separate contracts for launch services and TT&C ground support, rather than delegating this part of the work to the prime contractor. There is a growing trend, however, for customers to place so-called turn-key contracts with industry which include all the above services.

When following this commercial approach, it takes approximately 4 years from being awarded the contract to develop a flight-worthy satellite. Once built, it is submitted to the customer for final acceptance before launch.

Hardware Hierarchy

Before embarking on a discussion concerning model philosophy and test, it is useful to define units, subsystems and systems.

Units (sometimes called "equipments") are complex assemblies of components mounted in boxes or on substrates which can be moved around easily. Examples are solar arrays, batteries, regulators, encoders, decoders, microprocessors, attitude sensors, momentum wheels, deployment booms and individual antennae.

Subsystems are made up of units interconnected by electrical harnesses or structural members. They perform major and complex functions, and are not readily removable from the spacecraft. Chapters 5–13 describe the most common subsystems in geostationary applications satellites, namely structure, mechanisms, thermal control, power supply and conditioning, propulsion and orbit control, attitude control, TT&C, communications payload and meteorological payload. Variations in subsystem definitions exist. Attitude and orbit control functions are often combined into a single *Attitude & Orbit Control Subsystem* (AOCS), while communications antennae, as well as apogee and perigee motors, are sometimes considered subsystems in their own right.

System, in spacecraft engineering parlance, refers to the integrated satellite. Thus, "testing at system level" or "system integration" implies

Fig. 15.1. Hypothetical industrial consortium for building a meteorological satellite.

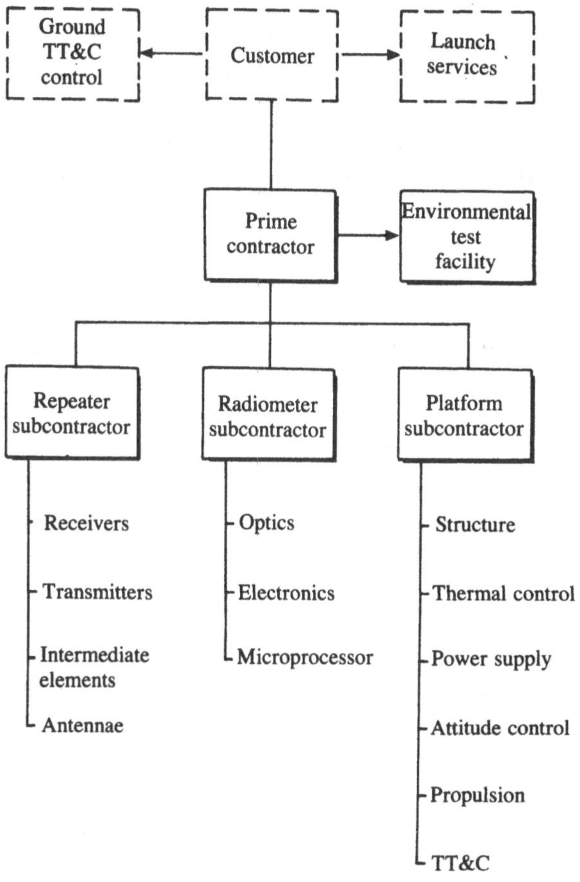

Ground TT&C control
Customer
Launch services
Prime contractor
Environmental test facility
Repeater subcontractor
Radiometer subcontractor
Platform subcontractor
Receivers
Transmitters
Intermediate elements
Antennae
Optics
Electronics
Microprocessor
Structure
Thermal control
Power supply
Attitude control
Propulsion
TT&C

activities involving the entire spacecraft, as opposed to components, units or subsystems.

Model Philosophy

In the early days of spacecraft technology, when almost every piece of satellite equipment was a new and untried design, disposable prototypes were built to gain technical confidence before embarking on the construction of expensive flight equipment. The prototypes were developed at several hardware levels, from components and units to subsystems and systems. The prototypes and the flight hardware were called *models*, and the way they were conceived became known as the model philosophy.

Nowadays the old model philosophy is still used as a programmatic baseline, but some of the old prototype concepts have been eliminated to save time and money. These deletions are justified on the grounds that many designs and production methods are inherited from previous programmes which have already successfully flown in space.

Let us now define classes of spacecraft hardware models. The definitions vary somewhat from one manufacturer to the next and should not be interpreted rigorously. The number of models built is determined by the level of design maturity.

A *breadboard* is a crude version of an electronic or electromechanical unit. The sole purpose of the breadboard is to give the designer an opportunity to validate functional concepts. Commercial rather than space qualified components are used. Printed circuit tracks are freely cut open or interconnected by soldering electric wires. The unit is not necessarily accommodated in any kind of housing, and there is no quality control.

A *brassboard*, also known as an "elegant breadboard", tries to emulate the final unit physically as well as functionally. It may include an authentic housing, and the component layout is tidier.

The following model categories exist at all hardware levels – unit, subsystem, as well as system. An *engineering model* strives to be physically and functionally identical to the ultimate flight hardware. The main differences are that commercial rather than flight standard components are used, and quality assurance rules are not enforced rigorously. Engineering model units are often built into an Engineering Model Spacecraft if the design is substantially different from any previous design. Only functional testing is performed.

A *qualification model* is built to full flight standard under rigorous quality control. The qualification model is submitted to environmental qualification testing as defined below. It is usually discarded after test but, because of its high cost, it is sometimes refurbished to become a flight model.

A *protoflight model* is subjected to qualification test levels but for reduced test durations. It is subsequently launched.

An *acceptance model* is synonymous with a *flight model*. Like the qualification model, it is built to full flight standard under strict quality control, but its exposure to environmental stresses before launch is limited to predicted worst-case flight values.

When a spacecraft configuration is contemplated which differs radically from already qualified or flight-proven designs, it is necessary to build a spacecraft *structure model* and submit it to static as well as dynamic load tests. In some programmes the structure model is subsequently refurbished to become a *thermal model*. This model is equipped with passive and active thermal control devices whose effectiveness is verified in space simulation tests. The structure or thermal model may eventually serve as the platform for an Engineering Model Spacecraft.

Inherited designs are a mixed blessing, for the net result is an uneven distribution across all subsystems of breadboards, brassboards, engineering models, qualification models, protoflight models and flight models. Managing such a variety of models and test programmes is not a trivial task, but it is usually justifiable in economic terms.

Assembly, Integration and Test (AIT)

Assembly means bringing all the units of a spacecraft together in order to begin mechanical and electrical integration. The complete spacecraft is subsequently tested. Testing at component and unit level, however, starts much earlier in the programme and is carried out more or less continuously. In addition to being subjected to a variety of development tests, each element of a spacecraft must at some point in time pass formal qualification and acceptance tests.

Qualification testing strives not only to verify functional performance, but also to produce confidence in the product by establishing design and performance margins over and above specifications. Margins are important to maximize the probability that the design will survive in the real flight environment. Because qualification test levels are more severe than expected flight levels (by 50–150%), the ability of the equipment to function properly after test is potentially degraded. Qualification hardware is therefore not normally used for flight.

While the aim of qualification testing is to establish margins, acceptance testing of flight equipment is purely for functional and workmanship verification purposes. Hence flight hardware is also tested, but more gently.

The objectives of testing spacecraft units vary according to their purpose, and the test methods depend on their design. In most satellite programmes the objectives are outlined in voluminous development and test plans, and the methods are spelled out in detailed test procedures. Given the multitude of unit-level tests, these documents take up several yards of shelf space. In the following we shall focus on testing at system level, i.e. on tests to which the satellite as a whole is subjected.

A typical AIT sequence is shown in Fig. 15.2. The satellite used in this example is made up of two halves, namely a payload module which carries all the electronics and antennae of the communications package, and a service module to accommodate the platform subsystems. The modular approach saves AIT time, since each half may be assembled, integrated and tested in parallel up to a point.

The payload module structure is dispatched to the payload sub-

Fig. 15.2. Typical assembly, integration and test sequence.

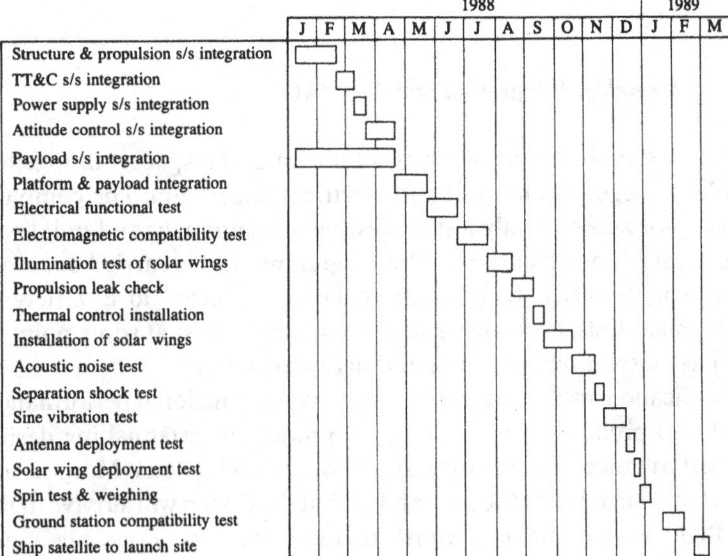

contractor. While the subcontractor is assembling antennae and transponder units, the prime contractor is building up the service module in logical steps. The first step is to assemble the bulk of the spacecraft by mounting the heavy propulsion subsystem onto the primary structure. An electrical wire harness is then installed, followed by the TT&C, power, and attitude control subsystems. Limited functional testing is carried out at each step to verify electrical compatibility between the various subsystems. Assembly of the complete spacecraft takes place as soon as the payload module is returned from the subcontractor.

To illustrate the logic behind the assembly sequence, let us imagine that Man was created by a Supreme Engineer. After subcontracting out the human head to a demi-god, he proceeded to mount lungs, kidneys and entrails on the skeleton. Before he could install and activate feeling, heart and muscles, he had to lay out blood vessels and a nervous system. Meanwhile the demi-god equipped the head with hearing, speech, vision and intelligence. Finally, the Supreme Engineer linked up the head and the body, and Man was ready for launch.

Once assembled, the spacecraft is ready for final integration and electrical characterization. Appendages such as booms and solar wings are installed and subjected to deployment tests in simulated weightlessness. Simulating a zero-g environment on earth is difficult, but some semblance of weightlessness is achieved by suspending appendages from gantries during deployment tests. Electrical characterization includes measuring solar array output, establishing a power consumption profile, and analysing the satellite's susceptibility to electromagnetic interference (EMI).

Having established that the satellite works on the ground, there is still no guarantee that it will do so in space. The biggest question mark is whether it will survive the violent ascent on top of the launch vehicle. In Chapter 4 (Fig.s 4.1–4.6) we showed flight load envelopes pertaining to some of the seven launch vehicles described in Chapter 1. Before a launch service agency will agree to fly a satellite, it requires evidence of tests having been conducted which demonstrate that the spacecraft will not break up due to acceleration, vibration or shock. The necessary proof is obtained by subjecting the complete spacecraft to a series of environmental tests, as explained in Table 15.1. Some of these tests are extremely onerous for the satellite, and any single failure can delay a programme for months.

Let us say that the satellite in our example is the first in a series of identical spacecraft. It is subjected to *protoflight* testing, which is a compromise between qualification and acceptance testing. The environmental test levels are the same as for full qualification, but the test durations are halved. The aim is to save time and money by obtaining an acceptable level of confidence in the design and performance margins without overstressing the satellite to the point where it must be discarded afterwards. All subsequent satellites in the series would be submitted to acceptance tests only, provided no design modifications are introduced which might invalidate their generic protoflight qualification status.

At the end of environmental testing, the satellite is given a thorough electrical performance check to verify that it has survived. After a final compatibility test with a representative TT&C ground station, the satellite is ready for packing and shipping to the launch site.

Table 15.1. *Environmental tests*

Type of test	Purpose
Acoustic vibration	To simulate the noise inside the rocket fairing
Random vibration	To simulate random vibrations entering the satellite through the attach fitting of the rocket
Sinusoidal vibration	To allow for all possible frequency contents. Pure sinusoids are especially useful for isolating resonances. Often a "sine sweep" is used instead of discrete frequencies
Shock	To simulate impulses on the satellite due to rocket engine cut-off and satellite separation
Solar simulation	To verify the adequacy of the satellite's thermal control by simulating solar illumination in vacuum
Thermal vacuum	To verify workmanship by exposing the satellite to hot and cold temperature cycles in vaccum
In addition to these environmental tests, it is customary to carry out:	
Spin and balance	To ensure that the satellite does not upset the flight stability of the launch vehicle, and that it will spin around the intended axis in orbit
Weighing	To ensure that the satellite mass corresponds to the launch vehicle lift-off mass capability
Mass properties determination	To establish moments of inertia for calibrating satellite attitude and orbit control in orbit
Electromagnetic compatibility	To ensure that onboard electronic oscillators do not interfere with each other through radiation or conduction

Test Facilities

A set of electrical ground support equipment (EGSE) and mechanical ground support equipment (MGSE) follows the satellite wherever it goes. The EGSE is an automated electrical check-out facility built around several miniprocessors. It powers up the spacecraft, switches on every conceivable configuration, provides artificial stimulus to dormant units (e.g. attitude sensors), and monitors the electrical performance. The MGSE consists of trolleys, tables, platforms and slings needed to transport and handle the spacecraft.

The environmental test facilities are huge, expensive installations located on the premises of major spacecraft manufacturers or, more commonly, of government-owned space research establishments. One of the largest and most modern test facilities in the world is located at the European Space Research and Technology Centre (ESA/ESTEC) in the Netherlands and is shown in Fig. 15.3.

Acoustic noise is generated by large horns which blare at the spacecraft inside a closed chamber. Random and sinusoidal vibration

Fig. 15.3. Large environmental test facilities. 1. Test area for small and medium size facilities. 2. Thermal and mechanical data handling facilities. 3. Users area. 4. Spacecraft check out room. 5. Test preparation area. 6. Physical properties machines. 7. Vibration systems multishaker (right) and 70 kN shaker (left). 8. Transport bay and airlock. 9. Electromagnetic compatibility (EMC) area EMC chamber (right) and small compact payload test range (left). 10. Large Space Simulator (LSS). Source: ESA.

tests are carried out by placing the satellite on top of a shaker. The satellite is spun and balanced on a spin table, while weighing takes place on a high-precision scale. Moments of inertia are determined by suspending the satellite from steel wires and measuring its oscillation period around various axes.

While launch vehicle vibrations can be simulated in a normal laboratory environment, space simulation requires the satellite to be installed in hermetically sealed chambers which are evacuated to 10^{-6} Torr. A *solar simulation chamber* is cooled by liquid nitrogen flowing through pipes embedded in the walls. At one end of the chamber, a set of high-intensity xenon lamps produce a parallel beam of simulated sunlight. The satellite is mounted on a *sting*, which is a pivot allowing the satellite to rotate and its attitude to be changed by remote control with respect to the artificial sunlight. A *thermal vacuum chamber* resembles a solar simulation chamber, except that the xenon lamps have been removed. Hot and cold thermal cycling is achieved by the use of temperature-controlled nitrogen flowing through wall pipes, as in the case of the solar simulation chamber, and by covering the walls with arrays of infrared lamps. The spacecraft is mounted on a spin table which turns slowly to ensure homogeneous heating and cooling.

Launch Campaign

After a satellite has successfully passed its environmental test programme, it is ready to be shipped to the launch site. The sheer bulk of a medium-size satellite and its associated EGSE and MGSE is enough to fill a Boeing-747 cargo aircraft. Upon arrival, the equipment is unloaded and transported on soft-suspension lorries to a spacecraft preparation facility (Fig. 15.4). The satellite is checked out once more before being moved to a different facility equipped with special safeguards for hazardous operation. Here the pyrotechnic devices are installed, and the propellant tanks are filled. The satellite is then mated to its solid propellant apogee motor or perigee motor, if applicable. The whole assembly is taken by road to the launch pad for installation onboard the rocket.

From now on, the satellite is out of the hands of the prime contractor and the customer. Its health can still be monitored through an *umbilical cable* which connects it with the EGSE on the ground, but any other intervention would require the satellite to be disembarked.

The countdown begins in earnest about 8 h before lift-off. The customer has the means to halt the countdown until a few seconds before ignition in the event that a satellite anomaly develops. While halted countdowns are common, most interruptions are caused by technical or operational anomalies in the launch vehicle or on the launch site, rather than by satellites.

Fig. 15.4. Ariane launch campaign sequence of events. Courtesy Arianespace.

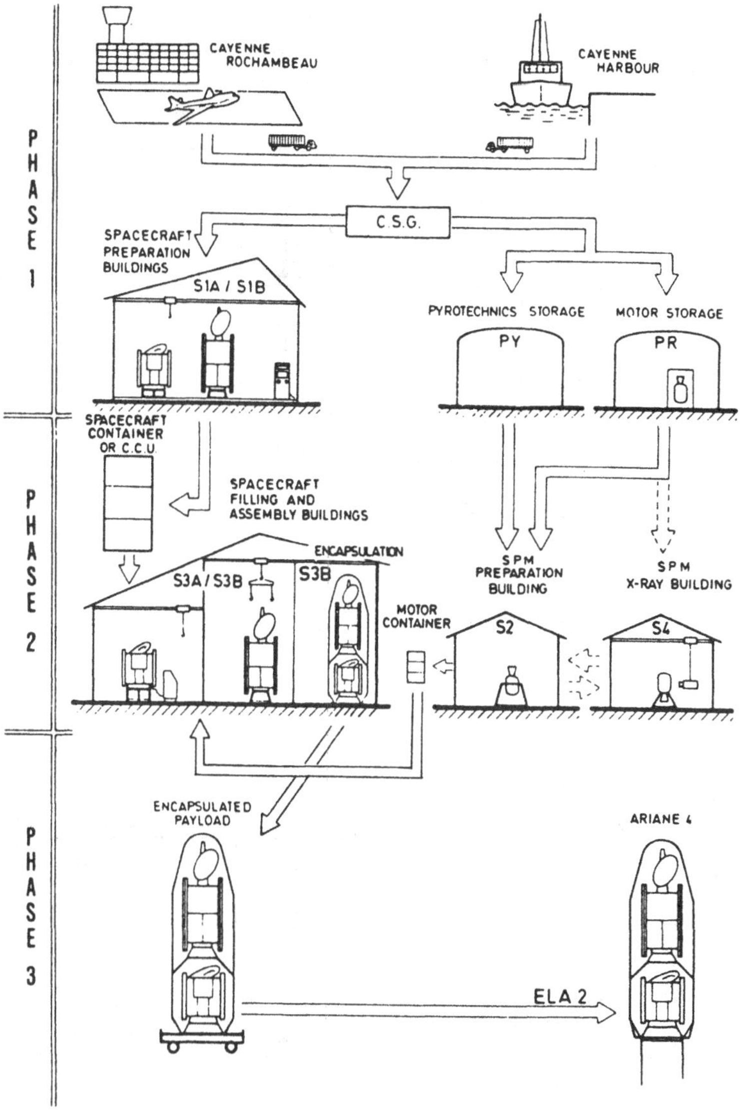

Fig. 15.5. Ariane launch campaign schedule. Courtesy Arianespace.

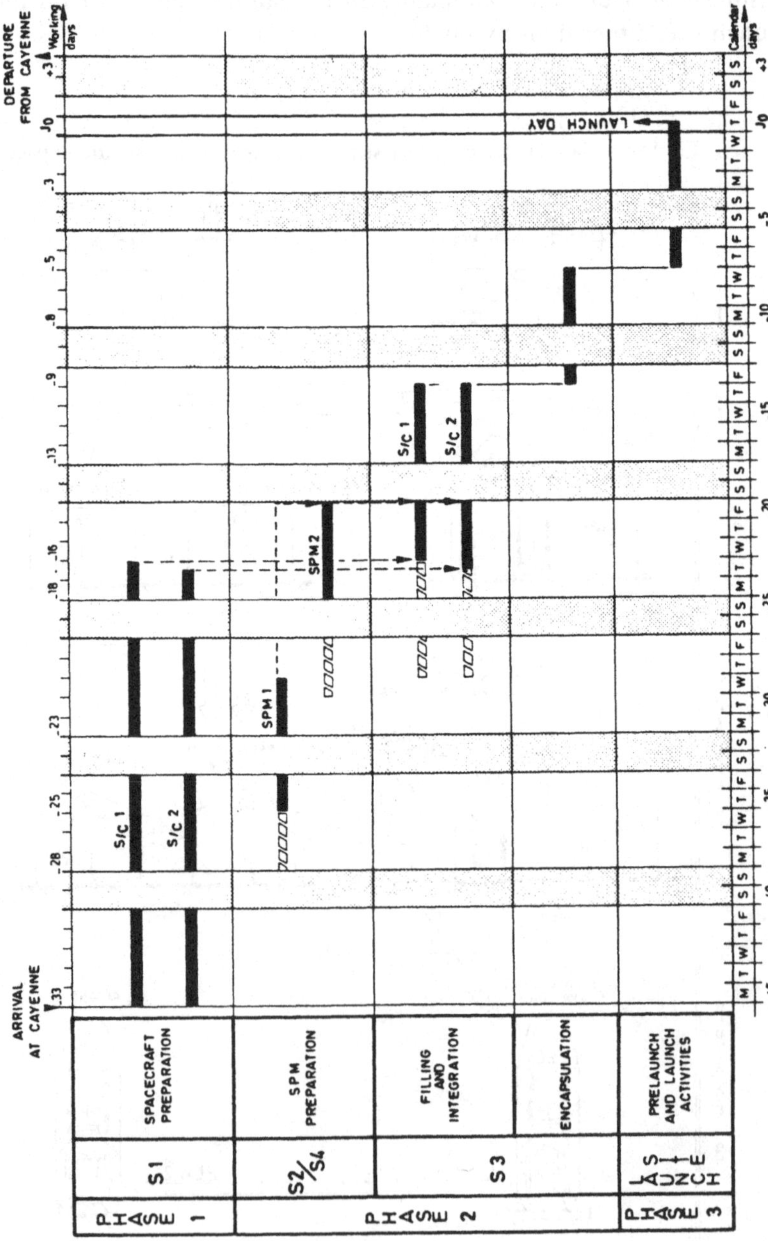

The Human Factor

A launch campaign involving an expendable rocket takes approximately 6 weeks from the arrival of the satellite until lift-off (Fig. 15.5). In the case of a shuttle launch the campaign is twice as long. Members of the customer's and the prime contractor's project teams work long hours to prepare the satellite and to manage the interfaces with the launch vehicle and with other satellites which are being readied in parallel. A launch campaign is a race against time, for whoever is late with his preparations runs the risk of holding up everybody else. If the end result is a launch delay, the culpable party becomes liable to indemnify the others at a rate of tens of thousands of dollars per day.

In the heat of a launch campaign, team members tend to reveal new and sometimes surprising qualities. A leader who is known for his composure at home might panic in critical situations on the launch site. Conversely, a normally unobtrusive character could develop a new assertiveness and step out into the limelight.

The mood of the team as a whole changes with time in a fairly predictable manner. Immediately after arrival, a feeling of exhilaration prevails while the satellite is unpacked and given its first post-shipment health examination. Gradually the long hours, the psychological pressure and the sometimes difficult working environment begin to wear people down. The lowest point is reached around the time when the satellite is handed over to the launch vehicle authority.

The team is then seized by a growing impatience to see the campaign come to a conclusion. As the launch date draws nearer, spirits are on their way up again. During terminal countdown, lift-off and ascent, the atmosphere is electric with apprehension. Once successful injection of the satellite has been confirmed, tension gives way to euphoria.

As the first telemetry signals are received on the ground, the responsibility for the satellite's well-being is removed from the project team and passes into the hands of operations personnel. The time has come for the team members to return home and plunge into new space programmes. Eventually, the satellite they took years to fashion, prime and expedite into orbit becomes but a distant memory.

Index

Printed in the United States
By Bookmasters